Wood Occupations

LEVEL 1

Peter Brett

Published in 2011 by:
Nelson Thornes Ltd
Delta Place
27 Bath Road
CHELTENHAM
GL53 7TH
United Kingdom

11 12 13 14 15 / 10 9 8 7 6 5 4 3 2 1

A catalogue record for this book is available from the British Library

ISBN 978 1 4085 0880 0

Cover photograph © Luis Albuquerque/iStockphoto
Page make-up by Wearset Ltd

Printed in Croatia by Zrinski

Every effort has been made to contact copyright holders and we apologise if any have been overlooked. Should copyright have been unwittingly infringed in this book, the owners should contact the publishers who will make corrections at reprint.

Acknowledgements
Bosch p87 (x3), p100TL, p100TM, p100BL; **Draper Tools** 103TM; **Fotolia** p29 (mearicon); **Hilti (Great Britain) Limited** p103BR; **Hitachi** p100BR, p100BM, p103TL; **iStockphoto** p1 (Wayne Pillinger), p13T (manley099), p13B (Steve Cukrov), p21 (×4), p49 (kkgas), p65 (Mark Wragg), p81 (flyfloor) p97 (Bogdan Shahanski); **Panasonic** p103BL; **Senco** p103TR; **Toolbank Marketing Services** p100TR

Contents

Introduction

Welcome to the Level 1 *Wood Occupations Course Companion*. It is literally a companion to support you throughout your course and record your progress!

This workbook-style book is designed to be used alongside **any** student book you are using. It is packed full of activities for you to complete in order to check your knowledge and reinforce the essential skills you need for this qualification.

Features of the *Course Companion* are:

Unit opener – a brief introduction to each unit

Key knowledge – the underpinning knowledge you must know is summarised at the beginning of each unit

Activities – a wide variety of learning activities are provided for you to complete in your *Course Companion*. Each activity is linked to one of the Personal, Learning and Thinking Skills to help you practice these fundamental skills:

 – Reflective Learner    – Self Manager

 – Creative Thinker   – Independent Enquirer

 – Teamworker   – Effective Participator

You will also notice additional icons that appear on different activities, which link to the following core skills and also to rights and responsibilities in the workplace:

 – Literacy

 – Numeracy

 – ICT

 – Employment, Rights and Responsibilities

Key terms – during your course you'll come across new words or new terms that you may not have heard before, so definitions for these have been provided.

Your questions answered – your expert author, Peter Brett, answers some burning questions you may have as you work through the units.

Quick Quiz – At the end of each unit you will find a multiple-choice quiz. Answering these will check that you have fully understood what you have learnt.

Good luck!

UNIT 1001

Safe working practices in construction

Health and safety forms an essential part of your daily working life. Ensuring that a site is as safe as it possibly can be is a shared responsibility between employers and the workforce.

Employers have a duty to create safe working conditions and provide the workforce with training that explains safety rules, procedures and regulations.

You, as part of the workforce, have a major contribution to make to site safety. You can do so by responding to safety instructions, complying with safety rules and developing the skills to identify potential safety hazards and reduce risks.

While at work you must comply with a wide range of legislation, regulations and supporting approved codes of practice (ACoP). Health and safety legislation is there to protect all persons at their place of work and other people from the risks occurring through work activities.

Key knowledge:
➤ health and safety regulations, roles and responsibilities
➤ accident and emergency procedures and how to report them
➤ hazards on construction sites
➤ health and hygiene in a construction environment
➤ how to handle materials and equipment safely
➤ basic working platforms
➤ how to work with electricity in a construction environment
➤ how to use personal protective equipment (PPE) correctly
➤ fire and emergency procedures
➤ safety signs and notices.

Health and safety regulations

Abbreviations and legislation

Health and safety **legislation**, people and organisations are often referred to by abbreviations or initials. You should be able to recognise these and know what they mean.

Your questions answered...

What can happen to me or my employer if health and safety legislation or approved codes of practice are not followed?

Failure to comply with the requirements of a piece of legislation or a set of **regulations** is a criminal offence that could result in prosecution. Failure to comply with an **ACoP** is not in itself an offence but, if a contravention of the associated regulations is alleged, failure to follow the ACoP will be accepted as evidence in a court of law.

ACTIVITY

Match each abbreviation concerned with health and safety legislation with its appropriate description:

PPER	The main legislation that covers health and safety at work.
	The body responsible for the enforcement of health and safety in the UK.
COSHH	Places a **duty** on employers, the self-employed and persons in control of premises to report to the HSE some accidents and incidents at work.
HASAWA	Requires employers to control exposure to hazardous substances in the workplace to prevent ill-health and protect both employees and others who may be exposed.
RIDDOR	
MHO	Requires all duty holders including the client, designers, building contractors, sub-contractors, site workers and others to play their part in improving onsite health and safety.
HSE	Requires employers and the self-employed to avoid the need to undertake manual handling operations that might create a **risk** of injury.
CDM	Requires employers to provide employees with any necessary personal equipment that is needed in order to carry out work safely.

Duties under health and safety legislation

Duties in legislation can be either '**absolute**' or have a qualifying term added, namely '**reasonably practicable**'.

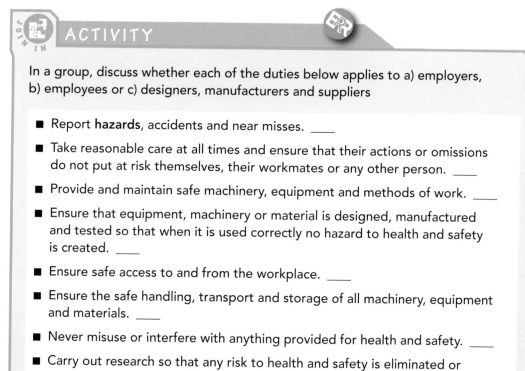

ACTIVITY

In a group, discuss whether each of the duties below applies to a) employers, b) employees or c) designers, manufacturers and suppliers

- Report **hazards**, accidents and near misses. ____

- Take reasonable care at all times and ensure that their actions or omissions do not put at risk themselves, their workmates or any other person. ____

- Provide and maintain safe machinery, equipment and methods of work. ____

- Ensure that equipment, machinery or material is designed, manufactured and tested so that when it is used correctly no hazard to health and safety is created. ____

- Ensure safe access to and from the workplace. ____

- Ensure the safe handling, transport and storage of all machinery, equipment and materials. ____

- Never misuse or interfere with anything provided for health and safety. ____

- Carry out research so that any risk to health and safety is eliminated or minimised as far as possible. ____

Health and safety facts

  ## ACTIVITY

Choose from the following terms to complete the facts about health and safety:

Induction   Risk assessment   Toolbox   Improvement   HSE   Prohibition   Safety officer

- _____ is the process of identifying hazards, assessing likelihood of **harm** and deciding on adequate control measures.

- An _____ notice issued by an _____ inspector requires employers to put right minor safety hazards within a specified period of time and a _____ notice requires work to stop immediately.

- All new workers to a site must undertake a site safety _____ before starting work.

- Deaths, major injuries and dangerous occurrences must be reported to the _____ without delay.

- _____ talk is the name given to the short talks often on safety topics, given by the site manager or _____.

ACTIVITY

List THREE of the HSE's main powers:

1.

2.

3.

ACTIVITY

In a group, discuss the purpose of toolbox talks and make a list of topics that could be discussed.

Accident and emergency procedures and how to report them

It is important that all accidents, incidents, emergencies and near misses in the workplace are reported, not only to comply with health and safety legislation but also to help prevent recurrences.

Construction is the UK's biggest industry and also one of the most dangerous. The HSE publish data based on the information collected under RIDDOR.

ACTIVITY

This pie chart shows the percentage distribution by severity of reportable accidents in the construction industry.

Undertake research in the library and on the web regarding the main causes of accidents for each of the three categories. Then, using a computer, record your findings in graphical form.

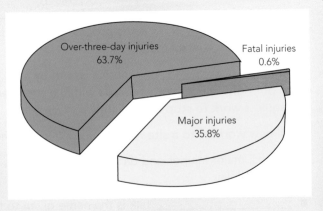

Over-three-day injuries
63.7%

Fatal injuries
0.6%

Major injuries
35.8%

ACTIVITY

Choose from the following words to fill in the blanks.

Hospital   Diseases   Deaths   Dangerous   Three   Amputations   Infections

Public   Work   Taken   Misses   Injury   Major   Accident

Under RIDDOR, employers, self-employed people and others in control of work premises have a legal duty to inform the HSE of the following events (reportable incidents):

- _____ arising out of or in connection with work.

- _____ injuries, which include most fractures, _____ , loss of sight, loss of consciousness, acute illness requiring medical treatment or any other injury involving a stay in _____ .

- Over-_____ -day injuries where a person is away from _____ or unable to perform their normal work role for more than three consecutive days.

- Injuries to members of the _____ or people not at work where they are _____ away from the scene of an _____ to hospital.

- Work-related _____ , which include poisonings, skin diseases, lung diseases, _____ , musculoskeletal disorders and vibration syndrome.

- _____ occurrences, which include the collapse of a crane, hoist, scaffolding or building, an explosion or fire, or the escape of any substance that is liable to cause a health hazard or major injury to any person.

- Certain near _____ , where something happens that does not result in an _____ but it could have done, may be classified as dangerous occurrences.

Your questions answered...

Can I go straight home in the event of an emergency rather than reporting to the assembly point?

No, you must follow the evacuation procedures, which are there for everyone's safety. If you don't report to your designated assembly point, other people may be placed at risk trying to find you.

Completing an accident record

Accident records are normally completed by the injured person. If this is not possible, then a witness or someone representing the injured person should complete it.

ACTIVITY

Read the following information:

Andy James is a Site Carpenter who works for BBS Construction and lives at 26 Fields Farm Lane, Long Eaton, Nottingham, NG10 2FF. He is currently working on the Station Road housing estate in Nottingham. At 4.30 p.m. on 18 August 2011, Andy tripped whilst carrying lengths of timber across uneven ground. Andy's reaction was to put out his right hand to break his fall. In doing so he released his grip on the timber, a length of which gave him a glancing blow on the forehead. As a result of this incident an ambulance was called to take Andy to the local hospital for an examination and X-rays. These tests confirmed that Andy had fractured his wrist and would need to wear a plaster cast for four to six weeks. The doctor stated that the blow to Andy's head appeared to have caused severe bruising. Andy was advised to go to his doctor if he suffered any headaches in future.

As a witness to Andy's accident, use this information to complete an accident record.

Report Number

ACCIDENT RECORD

Section 1: About the person who had the accident

Name

Address

Occupation

Section 2: About you, the person filling in this record

☛ If you did not have the accident, write your address and occupation.

Name

Address

Occupation

Section 3: About the accident

☛ Say when it happened.　　　　　Date　/　/　　Time

☛ Say where it happened. State which room or place.

☛ Say how the accident happened. Give the cause if you can.

☛ If the person who had the accident suffered an injury, say what it was.

☛ Please sign the record and date it.

Signature　　　　　Date　/　/　　Time

Section 4: For the employer only

☛ Complete this box if the accident is reportable under the Reporting of Injuries, Diseases and Dangerous Occurrences Regulations 1995 (RIDDOR).

How was it reported?

Date reported　/　/　　　　Signature

Accident and emergency procedures

ACTIVITY

In a small group discuss why it is important to make a record of an accident and why 'near misses' should be reported.

ACTIVITY

State THREE actions that should be taken by the first person to arrive on the scene of an accident:

1.

2.

3.

Your questions answered...

Why can't my employer keep tablets and medication in the first-aid box?

HSE guidance states that the giving of tablets and medication does not form part of first-aid treatment; therefore they should not be kept in a first-aid kit.

This is because without knowing a person's medical history the use of tablets and other medication can do more harm than good. In addition, if someone did have an allergic reaction as a result of a first-aider administering medication, the first-aider could be sued.

Hazards on construction sites

Carrying out a risk assessment

A risk assessment is a careful examination of the workplace to identify hazards and put measures in place to control the risk of an accident occurring.

ACTIVITY

Consider the five steps to risk assessment:

- **Step 1.** Look for the hazards.

- **Step 2.** Decide who might be at risk and how.

- **Step 3.** Evaluate the risks and decide on the action to be taken.

- **Step 4.** Record the findings.

- **Step 5.** Review the findings.

Look around your place of work or your college practical area and identify a hazard. Complete this risk assessment form.

BBS Construction Services

RISK ASSESSMENT

Activity covered by assessment: _____

Location of activity: _____

Persons involved: _____

Dates of assessment: _____

Tick appropriate box ✓

- Does the activity involve a potential risk?   YES ☐   NO ☐

- If YES can the activity be avoided?   YES ☐   NO ☐

- If NO what is the level of risk?   LOW ☐   MEDIUM ☐   HIGH ☐

- What remedial action can be taken to control or protect against the risk?

1 _____

2 _____

3 _____

4 _____

5 _____

MANAGEMENT SUMMARY:

Priority for action: LOW ☐   MEDIUM ☐   HIGH ☐

Action to be taken: _____

Date action to be taken by: _____

Date for reassessment: _____

Assessor's name and signature: _____

ASSESS THE RISK - PUT IN CONTROLS - CHECK THEY WORK

Method statements

A method statement is a key safety document, which takes the information about potential risks from a risk assessment and combines them with a job specification.

ACTIVITY

Read this employer's safety method statement concerning the use of MDF panel products. In your own words, describe the risks involved and the correct work procedures to be followed.

BBS: Shopfitting Services
33 Stafford Thorne Street
Nottingham NG22 3RD
Tel. 0115 94000

SAFETY METHOD STATEMENT

Process:

The remanufacture of MDF panel products. During this process a fine airborne dust is produced. This may cause skin, eye, nose and throat irritation. There is also a risk of explosion. The company has controls in place to minimise any risk. However, for your own safety and the protection of others, you must play your part by observing the following requirements.

General Requirements:

At all times, observe the following safety method statements and the training you have received from the company.
- Manual Handling
- Use of Woodworking Machines
- Use of Powered Hand Tools
- General Housekeeping

Specific Requirements:

- When handling MDF, always wear gloves or barrier cream as appropriate. Barrier cream should be replenished after washing.
- When sawing, drilling, routing or sanding MDF, always use the dust extraction equipment and wear dust masks and eye protection.
- Always brush down and wash thoroughly to remove all dust, before eating, drinking, smoking, going to the toilet and finally at the end of the shift.
- Do not smoke outside the designated areas.
- If you suffer from skin irritation or other personal discomfort, seek first-aid treatment or consult the nurse.

IF IN DOUBT ASK

Hazards on construction sites

ACTIVITY

List THREE hazards that are associated with chemical spills in the workplace:

1.

2.

3.

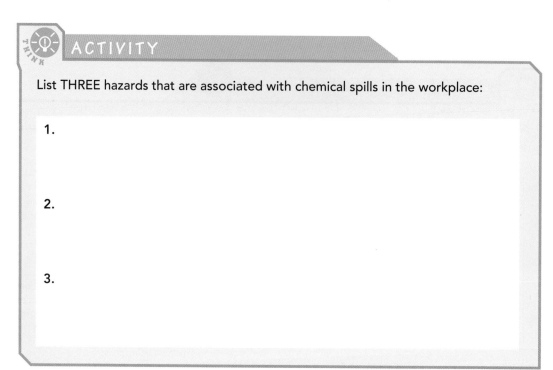

ACTIVITY

Explain THREE safety points concerning the storage of flammable liquids and gases:

1.

2.

3.

ACTIVITY

You are working on a construction site and have been asked to maintain good housekeeping in the workplace. Consider what is meant by this. Using a computer, make a list of what this requires and also any action that should be taken if a hazard is identified in the workplace.

Health and hygiene in a construction environment

Provision of welfare facilities

Welfare facilities include toilets, changing rooms, somewhere to wash, eat and rest and the provision of drinking water. Employers must provide these facilities to ensure health and safety in the workplace.

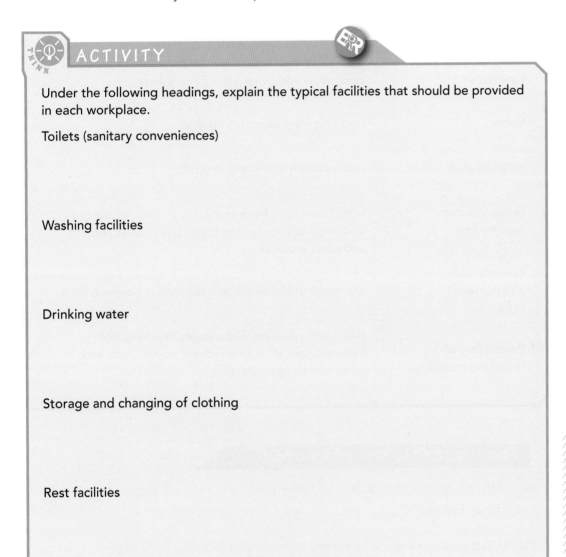

ACTIVITY

Under the following headings, explain the typical facilities that should be provided in each workplace.

Toilets (sanitary conveniences)

Washing facilities

Drinking water

Storage and changing of clothing

Rest facilities

Your questions answered...

Why have I been told at my site induction not to use mobile phones, radios and music players during working hours?

Their use can cause a loss of concentration, as well as interfere with general communications and make emergency warnings, etc. harder to hear.

Health problems associated with construction

The use of chemicals and other hazardous substances are a major risk to people's health on construction sites and other workplaces.

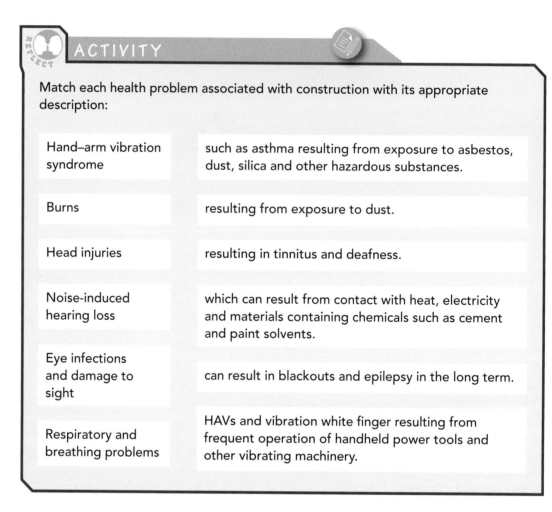

ACTIVITY

Match each health problem associated with construction with its appropriate description:

Hand–arm vibration syndrome	such as asthma resulting from exposure to asbestos, dust, silica and other hazardous substances.
Burns	resulting from exposure to dust.
Head injuries	resulting in tinnitus and deafness.
Noise-induced hearing loss	which can result from contact with heat, electricity and materials containing chemicals such as cement and paint solvents.
Eye infections and damage to sight	can result in blackouts and epilepsy in the long term.
Respiratory and breathing problems	HAVs and vibration white finger resulting from frequent operation of handheld power tools and other vibrating machinery.

Your questions answered...

Is it OK to go to the pub and have just one pint of beer during a lunchtime break?

No, as just one drink leads to slower reflexes and reduced concentration levels, which puts not only you but also everyone in the workplace at an increased risk of danger. In addition, it is almost certainly against your employer's safety policy, which you would have been advised of during your induction, and doing this could result in your dismissal.

ACTIVITY

Produce a list of FOUR precautionary control measures that may be taken to prevent or reduce exposure to hazardous substances:

1.

2.

3.

4.

Handle materials and equipment safely

Mechanical and manual handling aids

Manual handling should only be undertaken as the last resort after considering all other mechanical means and manual aids available.

ACTIVITY

Look at the photographs in the margin, which show some mechanical and manual handling aids. Name each item and describe a situation where each would be suitable to use:

A

B

A

B

13

How much can I safely lift and when should I seek help?

Although the maximum load that one person should move is 25 kg, it is dependent on your stature and competence. Two or more persons are required to move heavier, larger or awkward shaped items to reduce the risk of injury.

Mechanical versus manual handling

 ACTIVITY

Either in a small group or on your own, consider the following real-life problem or scenario:

> Kevin and John work as a pair and have been employed by a sub-contract carpentry firm for the last six months working on a large housing project. When they arrived for work one morning, there was a lorry containing long floor **joists** at the site entrance. They decided to offload the joists by hand as the plant driver had not yet arrived.
>
> While collecting one of the timber joists from the vehicle's driver, John lost his footing on the uneven ground. He twisted quickly in an attempt to maintain his balance, but ended up in a heap on the ground, with the joist giving him a glancing blow to the head. An ambulance was called to take John to hospital for emergency treatment. John was off work for three weeks while his sprained back settled down, but he still gets a reoccurring back twinge if he moves quickly and occasional bad headaches that the doctor has put down to the blow to his head.

- Whose fault was the accident?

- What actions do you think Kevin and John should have taken to avoid this accident?

key terms

Joist: one of a series of parallel beams that span the gap between walls in suspended floors and roofs, to support floor, ceiling and flat roof surfaces.

I've been told that manual handling is a major cause of workplace musculoskeletal disorders. What are they and how do they develop?

Musculoskeletal disorders or MSDs are those disorders that affect the joints, muscles, tendons, ligaments and nerves and include repetitive strain injuries. Most MSDs develop over time from years of moving heavy items, awkward or bulky shapes or using poor procedures and postures.

Safe lifting procedures

Where avoidance of manual handling is not reasonably practicable, the adoption of safe lifting procedures will reduce the potential risk of injury.

ACTIVITY

Draw lines between the appropriate boxes to connect the two parts of the practical tips to be followed when lifting materials and equipment:

Get a good grip:	On a suitable platform, ensuring your hands and fingers will not be trapped, before sliding the load into position. Where the load has to be lowered to the floor level, you should adopt a similar position to that used when lifting.
Think first:	Always lift with your back straight, elbows tucked in, knees bent and feet slightly apart.
Place load down first:	Can I use an aid? Do I need help? What PPE is appropriate? Are there splinters, nails and sharp or jagged edges on the items to be moved?
Take up the correct position:	Avoiding any twisting and leaning, keeping the load close to your body and look straight ahead rather than down at the load.
Move smoothly:	Using your leg muscles and not your back.
Lift the load:	Hands should be placed under the load and, on lifting, the load should be hugged as close as possible to your body.

Handle materials and equipment

ACTIVITY

State THREE checks to be made before manually moving materials from storage to the workplace to ensure the intended route is clear and safe:

1.

2.

3.

ACTIVITY

Number the following action points, which are aimed at minimising the amount of waste material in construction, to identify the correct order of priority:

- ☐ Dispose of waste in a landfill site.
- ☐ Eliminate waste wherever possible.
- ☐ Reuse materials that are potential waste.
- ☐ Reduce the amount of waste created.
- ☐ Recycle waste materials wherever possible.

Basic working platforms

Each year in the construction industry, numerous falls take place as a result of the use of working platforms and access equipment such as scaffolding and ladders. It is therefore of major importance that the risks of working at height are assessed and appropriate work equipment is selected and used.

ACTIVITY

Look at the illustrations in the margin, which show a range of working platforms and access equipment. Name each item and describe a situation where each would be suitable to use:

A

B

C

Your questions answered...

Why can't I use a milk crate, saw stool or chair as a hop-up to gain access to low level work, which is just out of reach?

These should never be used as they are not designed for this purpose and may overturn or collapse completely, resulting in injury.

Stepladders

Stepladders are used mainly for short duration internal work on firm, flat surfaces and should be inspected on a daily basis by the person intending to use them.

 ACTIVITY

Produce a pre-use list of checks to be made before using stepladders:

Scaffolds

Scaffolding is a temporary structure that is used in order to carry out certain building operations at height. Tubular scaffolding and fittings is the most common type of scaffolding used in the construction industry to provide a safe means of access to heights and a safe working platform.

 ACTIVITY

Use the internet to research types of tubular scaffolding and complete the following tasks:

1. Name different types of tubular scaffolding.

2. State THREE requirements for a ladder that is used to provide access to the scaffold's working platform.

    1.

    2.

    3.

Is it OK if I temporarily remove part of a scaffold in order to make my work easier?

No, you must never remove any part of a scaffold as it can weaken it and you may be responsible for its total collapse. Scaffolding must only be erected, altered or dismantled by a trained, competent 'carded scaffolder'.

What is a 'carded scaffolder'?

A 'carded scaffolder' is a person who holds a recognised skills card or certificate showing that they have been trained and assessed as being competent in the erection, alteration, dismantling and inspection of tubular scaffolding.

Basic working platforms

 ACTIVITY

List FOUR general visual checks that should be made before using a scaffold:

1.

2.

3.

4.

 ACTIVITY

Using the internet or a textbook, look up the *Work at Height Regulations*. State THREE duties they place on employers and give reasons why:

1.

2.

3.

ACTIVITY

In a group consider these questions:

- What determines the maximum safe working height of a mobile tower scaffold?
- What is the maximum recommended load that can be carried up or down a ladder?
- What is the purpose of **handrails**, mesh guards and toe boards that are fitted to working platforms?

Working with electricity in a construction environment

Colour codes for different voltages

Electrical supply cables and industrial shielded plugs and sockets are colour coded to show their purpose and the voltage they are carrying.

ACTIVITY

Give the voltage and intended use of each of the colour codes:

Red:

Blue:

Yellow:

Your questions answered...

I've been told that my drill's battery charger must be PAT tested and receive a pass certificate before I can use it in the workplace. What is a PAT test and how often must it be carried out?

PAT is the abbreviation used for portable appliance testing. Under the Electricity at Work Regulations all portable electrical appliances used in the workplace have to be tested by a competent person at regular intervals, to ensure they are safe to use. PAT testing is normally carried out on an annual basis, but can be varied depending on the frequency of use.

Working with electricity

ACTIVITY

In a small group, consider the following real-life problem or scenario:

Sanjay is a recently qualified carpenter who is undertaking refurbishment work for a specialist contractor.

While he was working on battening out walls and fixing a plasterboard lining, he had to cut out holes for electrical sockets and pull the cables through.

The electrician had taped over the exposed ends of the cables for protection; all was going well until there was a loud bang and sparks as Sanjay pulled the cables though the last box. Sanjay received an electric shock and burns to his fingers, which required hospital treatment and time off work.

- What do you think was the cause of Sanjay's accident?

- Whose fault was it?

- What precautions should Sanjay take in the future when working with or near electrical equipment?

ACTIVITY

1. What voltage is specified when using portable electrical equipment on building sites?

2. Name the item of equipment that can be used to operate reduced voltage electrical equipment from a mains supply.

3. Explain why using reduced voltage electrical equipment is safer than using just a mains supply.

Using personal protective equipment (PPE) correctly

PPE

PPE is the equipment and clothes a worker needs to wear, or use, to protect them from risks to their health and safety that cannot be eliminated or adequately controlled in other ways. Not using PPE, or using inappropriate or damaged PPE, can cause injuries, ill-health and industrial diseases.

When choosing items of PPE, you should ask yourself the following questions:

➤ Is it the right item for the working conditions and the risks involved?

➤ Will it help to control the risks without adding to them?

➤ Can it be adjusted so that it fits correctly and is comfortable to use?

➤ Will you still be able to carry out your work task properly while wearing the item?

➤ If you need to use more than one item of PPE at the same time, are they compatible?

➤ Is the item well maintained and in good working order?

Types of PPE

ACTIVITY

Look at the photographs. Name each item and state a situation where it is required:

Is it safe to continue wearing a damaged item of PPE until it can be replaced?

No, as it will not provide adequate protection. Any items of PPE that have been damaged or show signs of wear must not be worn and should be replaced immediately before any further work is undertaken.

ACTIVITY

In the table below, explain for each of the following risks the appropriate actions to be followed and the PPE required:

Risk	Action and PPE
Skin cancer	
Cuts and infections	
Head injury	
Leptospirosis	
Hearing damage	

ACTIVITY

Name the items of PPE that are considered mandatory and should be worn onsite at all times:

ACTIVITY

You are working on a construction site and you notice someone wearing a dust mask but there are toxic fumes present. Why is this an incorrect use of PPE? What would you say to the person?

Fire and emergency procedures

Elements essential for a fire

The three essential elements that are required for a fire to ignite and burn are fuel, oxygen and heat or an ignition source.

 ACTIVITY

1. What is the name given to the illustration?

2. Explain what it represents.

3. What would be the result of removing one of the elements?

Methods of fire prevention

Employers should undertake a risk assessment of the workplace to establish the type and number of fire extinguishers required in the event of an emergency. Employers also have a duty to undertake certain actions in order to eliminate or reduce the risks in the workplace.

 ACTIVITY

List THREE potential fire hazards that should be taken into consideration during a risk assessment:

1.

2.

3.

Procedures in emergency situations

Emergencies are situations or events that require immediate action and it is essential that individuals are aware of the procedures and responsibilities for dealing with them.

ACTIVITY

List FIVE actions that members of the workforce should take in the event of a fire or other emergency situation arising:

1.

2.

3.

4.

5.

Types of fire extinguishers

There are various fire extinguishers and fire blankets available for tackling a fire in its early stages. It is essential that the correct type is selected, as using the wrong type can make the fire spread or cause injury to the user.

ACTIVITY

Look up the types of fire extinguisher on the web or in a textbook. Using the table below, state the content of each and what type of fire they are suitable to be used on.

Label colour	Content	Type of fire

Your questions answered...

Is it OK to go back into the building and help put out a fire after reporting to the assembly point?

No, the use of fire extinguishers should only be considered for small fires in their early stages or where the fire is blocking your escape route. Under no circumstances should you re-enter the site or building until the nominated person authorises it.

Safety signs and notices

Types of safety sign

A range of safety signs can be seen displayed around the workplace. Each has a designated shape, colour and symbol or pictogram to ensure that health and safety information is presented to employees in a consistent, standard way.

 ACTIVITY

Look at the illustrations in the margin, which show the basic shapes and colours used for safety signs. Name and state the purpose of each sign:

Use of supplementary text safety signs

Although the purpose of a safety sign is identified by its designated shape, colour and symbol or pictogram, they are often supported by supplementary text to provide additional information and instruction.

THINK ACTIVITY

Add appropriate supplementary text for each sign as an aid to understanding:

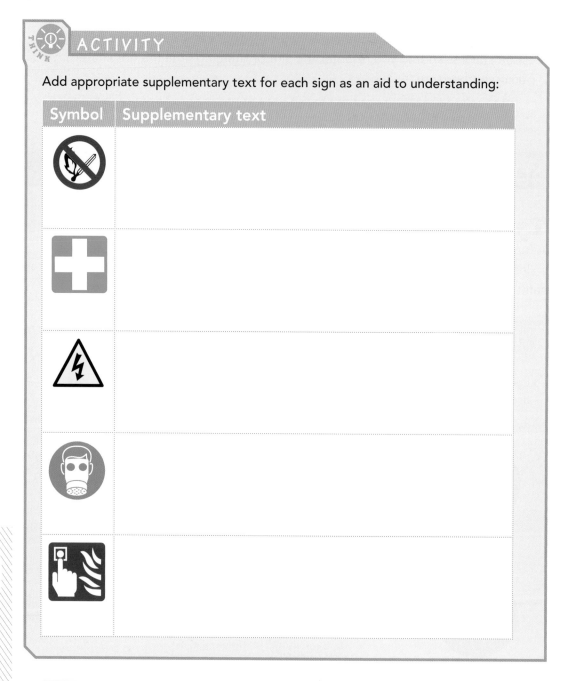

Symbol	Supplementary text

Your questions answered...

What are hand signals used for?

A hand signal can be used to direct hazardous operations, such as crane and vehicle manoeuvres. Anyone giving hand signals must be competent, wear distinctive brightly-coloured (hi-viz) clothing and use the standard arm and hand movements.

Spot the hazards

 ACTIVITY

Look at the illustration of an unsafe building site and identify safety hazards, breaches of regulations and general unsafe practices. Circle each one and number it – there are at least 10 but you may find more.

Describe each hazard in the following list.

1. ....
   ....
   ....

2. ....
   ....
   ....

3. ....
   ....
   ....

4. ....
   ....
   ....

5. ....
   ....
   ....

6. ....

7. ....

8. ....

9. ....

10. ....

QUICK QUIZ How much do you know about safe working practices in construction?

1. Which one of the following abbreviations refers to the main legislation that deals with health and safety in all workplaces?
   a. COSHH
   b. HASAWA
   c. PUWER
   d. RIDDOR

2. Why is it important to report a near miss, even if no one is hurt?
   a. it should be recorded in the accident book
   b. lessons can be learned, which help to prevent future accidents
   c. you may be able to get compensation
   d. someone may have to be disciplined

3. What is the purpose of a safety sign that has a red circle and diagonal line on a white background?
   a. identifies fire equipment or where to find it
   b. indicates something that you must not do
   c. indicates something that you must do
   d. warns of a specific hazard

4. What is the principal cause of fatal accidents on construction sites?
   a. contact with moving machinery
   b. contact with electricity
   c. falls from a height
   d. struck by a falling object

5. Under the Health and Safety at Work Act, which one of the following is an employee's duty?
   a. ensure that equipment is safely designed
   b. prepare a safety policy
   c. use the equipment and safeguards provided
   d. provide and maintain a safe working environment

6. Which one of the following is not an employer's responsibility under the Health and Safety at Work Act?
   a. provide employees with the necessary information, instruction, training and supervision to ensure safe working
   b. provide employees with safe transportation to and from work
   c. provide and maintain a safe working environment
   d. ensure safe access to and from the workplace

7. Legislation is:
   a. something that only applies to employers
   b. a guide to explain best health and safety practices
   c. a law or set of laws that must be followed by all
   d. an approved code of practice that gives explanations of the law

8. Which one of the following regulations deals with the control of hazardous substances?
   a. PUWER
   b. COSHH
   c. RIDDOR
   d. PPER

9. A fire extinguisher containing water can be used on:
   a. Class A fires
   b. Class B fires
   c. Class C fires
   d. Class E fires

10. What colour is used in the workplace to denote that the equipment uses a 110 volt supply?
    a. red
    b. blue
    c. yellow
    d. black

UNIT 1002

Information, quantities and communicating with others

Working in the construction industry involves working with and interpreting a wide range of information to suit the needs of a project, including determining the material quantities required for a particular task. At the same time as working with information, you will be working within a team and **communicating** with other people.

Key knowledge:
➤ interpreting **building information**.
➤ determining quantities of materials.
➤ relaying information in the workplace.

key terms

Building information: the key contact documents produced by the clients, such as the working drawings, bill of quantities, specification, schedules, specification and conditions of contract. In addition it also refers to the work programmes produced by a contractor and any manufacturers' information.

Communicating: the exchange of thoughts, messages and information, by speech, body language, written or printed information and telecommunications.

Interpreting building information

Drawing symbols

ACTIVITY

Symbols and abbreviations are used on drawings instead of words in order to allow the maximum amount of information to be included in a concise way, without cluttering the sheet.

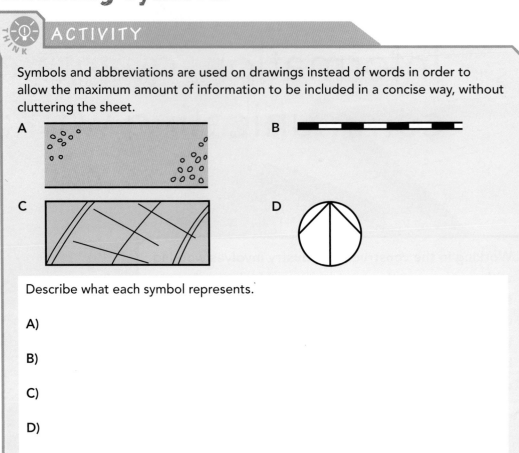

Describe what each symbol represents.

A)

B)

C)

D)

Scales

Scales are ratios that permit measurements on a drawing or model to relate to the real dimensions of the actual job.

ACTIVITY

Complete the table to show the actual size. It's simply a matter of multiplying the scale measurement by the scale ratio.

Scale ratio of join	Size drawn	Actual size
1:1	100 mm	100 mm
1:5	250 mm	1250 mm
1:10	100 mm	1000 mm
1:20	75 mm	
1:50	125 mm	
1:200	150 mm	
1:200	125 mm	
1:1250	25 mm	
1:2500	50 m	

Scale rules

 ACTIVITY

Scale rules are used to save having to calculate the actual size represented on drawings. They have a series of marks engraved on them so dimensions can be taken directly from the rule.

Select the appropriate scale rule and mark on it lines to represent the following measurements:

1) 8 m to a scale of 1:100.

2) 1300 mm to a scale of 1:20.

3) 360 mm to a scale of 1:5.

4) 96 m to a scale of 1:1250.

Use of scale rule

Scale rules can be useful when reading drawings and determining material requirements. However, preference should always be given to written dimensions shown on a drawing, as mistakes can be made.

ACTIVITY

The outline ground-floor plan of a detached house has been drawn to a scale of 1:100. Use your scale rule to measure the lengths. Complete the table.

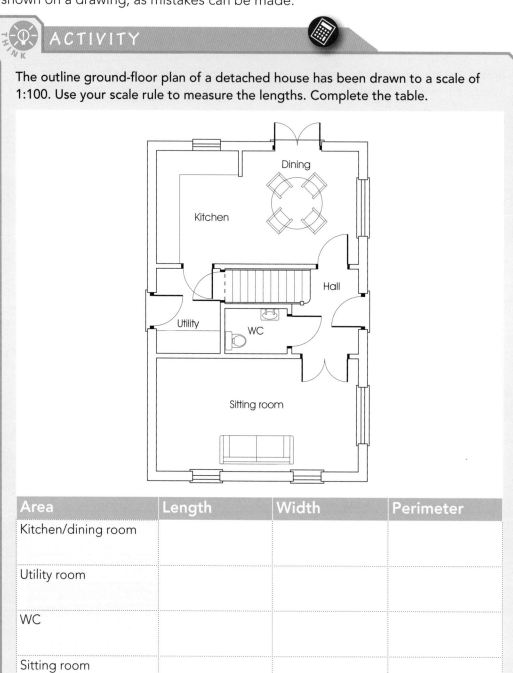

Area	Length	Width	Perimeter
Kitchen/dining room			
Utility room			
WC			
Sitting room			

Working drawings

These are scale drawings showing the plans, elevations, sections, details and locality of an existing or proposed building. They may also include figured dimensions, printed notes and a title panel, which identifies and provides information about the drawing.

ACTIVITY

Look at the illustrations, which show a range of working drawings. Name each type of drawing.

A

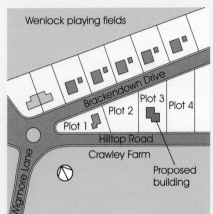

Wenlock playing fields

Brackendown Drive

Plot 1
Plot 2
Plot 3
Plot 4

Hilltop Road

Crawley Farm

Wigmore Lane

Proposed building

B

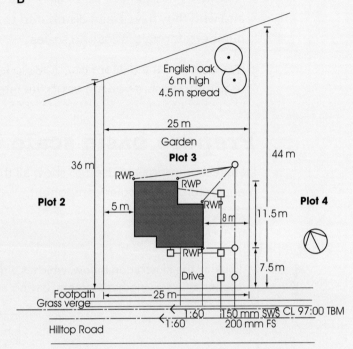

English oak
6 m high
4.5 m spread

25 m

Garden

Plot 3

44 m

36 m

RWP

RWP

Plot 2

5 m

RWP

8 m

11.5 m

Plot 4

7.5 m

RWP

Drive

Footpath

25 m

Grass verge

1:60   150 mm SWS   CL 97:00 TBM
1:60   200 mm FS

Hilltop Road

C

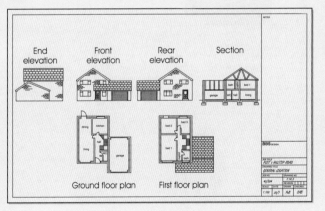

| End elevation | Front elevation | Rear elevation | Section |

Ground floor plan   First floor plan

D

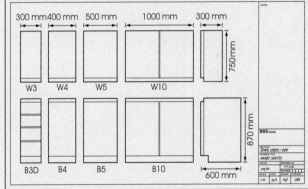

300 mm 400 mm 500 mm 1000 mm 300 mm

750 mm

W3   W4   W5   W10

870 mm

B3D   B4   B5   B10   600 mm

A)

B)

C)

D)

What is a drawing register and when is it used?

A drawing register is a list of all of the drawings relevant to a particular job and who they have been distributed to, as well as drawing numbers, revision numbers, formats, sizes and scales.

It is essential that before any drawing is used, it is checked against the drawing register to ensure you are using the latest version.

Prepare basic scale drawings

Detailed drawings are used to show all the information required in order to manufacture a particular component.

 ACTIVITY

Look at the illustration below, which is a detailed drawing showing the vertical section of a serving hatch frame, having an overall size of 1145 × 765 mm.

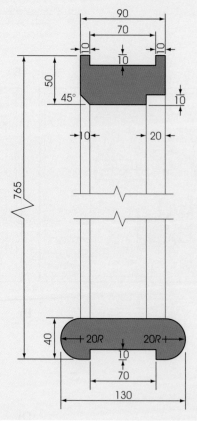

On a separate sheet of paper, produce a 1:2 scale drawing of the vertical section and add a horizontal section. Label the component parts and add the appropriate hatching to show that the hatch is made from hardwood.

Interpreting building information

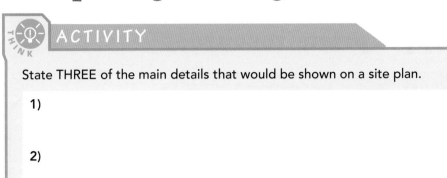

ACTIVITY

State THREE of the main details that would be shown on a site plan.

1)

2)

3)

ACTIVITY

List THREE factors that could result in delays to a contract programme.

1)

2)

3)

ACTIVITY

Complete the table to show what the abbreviations mean.

Abbreviation	What does it stand for?
RWP	
bldg	
DPC	
DPM	
jst	

Determining quantities of materials

Select appropriate resources

ACTIVITY

Name FIVE items of equipment that you may select to help you with calculations.

1)

2)

3)

4)

5)

ACTIVITY

Name FIVE written or printed forms of information that you might have to refer to when doing calculations.

1)

2)

3)

4)

5)

Area and perimeter of basic shapes

The **area** and perimeter of basic shapes can be found by using standard **formulae**.

 ACTIVITY

The table contains the formulae that can be used to calculate the area and perimeter of a range of common shapes.

Complete the table to show which of the following shapes the different formulae apply to. Add a sketch of each shape, showing how the formula applies

➤ rectangle
➤ triangle
➤ circle
➤ square.

Shape	Area equals	Perimeter equals
	AA	4A
	LB	2(L + B)
	$\dfrac{BH}{2}$	A + B + C
	πR^2	πD or $2\pi R$

Why are centimetres not normally used as a measure of length in the construction industry?

Centimetres are not normally used to avoid confusion. The preferred units of length are metres and millimetres. The length of a room can be stated as either 4255 or 4.255, and it is obvious from what is being described whether the dimension is in metres or millimetres.

Estimating material quantities

Where metres and millimetres are contained in the same problem, you should first either convert the millimetres into a decimal part of a metre or, alternatively, convert all units to millimetres.

ACTIVITY

The illustration shows a simple, dimensioned sketch of a room in a domestic house.

2000 mm

3600 mm

900 mm door opening

2 m

1.6 m

2.5 m

1) Calculate the area of the room and determine the number of 600 × 2400 mm chipboard sheets required to cover it.

2) Calculate the amount of tongue-and-groove matchboarding required to clad the 3600 mm-long wall up to a height of 2.1 m, if the boards have a covering width of 95 mm.

3) Determine the total length of skirting required for the room.

Complex shapes

The area of complex shapes can be calculated by breaking them into a number of recognisable shapes and solving each one in turn.

 ACTIVITY

Calculate the area and perimeter of the shape.

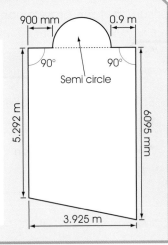

Determine quantities of materials

 ACTIVITY

1) Convert the following dimensions into metres:

   A) 2864 mm   **B)** 785 mm

   C) 34,700 mm   **D)** 72 mm

2) What is the width of a rectangular room that is 4.5 m long and has a total perimeter of 15.5 m?

3) Explain the relationship between the radius, diameter and circumference of a circle.

4) Deter mine the total run, in metres, of 150 mm-wide soffit board that may be cut from a 1220 × 2440 mm sheet of plywood?

5) Calculate the triangular area of a gable end wall that is 3.6 m wide and 1.4 m high.

Relaying information in the workplace

Methods of communication

Communication is a means of passing information between people. Effective communication is essential between all members of the building team in order to ensure that the project operates smoothly, runs on time and finishes within budget.

ACTIVITY

State TWO examples of each method of communication being used in the workplace.

Verbal communication

1)

2)

Body language

1)

2)

Written or printed information

1)

2)

Visual and graphic information

1)

2)

Telecommunications or electronic communications

1)

2)

Positive and negative communication

Positive communication can boost the motivation of the recipient and create a sense of achievement. Negative communication can damage people's confidence, reduce motivation and lead to feelings of despair.

ACTIVITY

Draw a line between the appropriate boxes to connect the key words and statements concerning positive communication.

Key words	Statements
Be personal	On occasion, it may be necessary to expose and tackle problem situations in order to resolve them, but you should accentuate the positive to maintain people's desire to achieve and assist. For example, instead of saying 'you made a right mess of this job', consider 'on this occasion your work fell below your normal high standards'.
Think	Help recipients understand what's being communicated. Poor handwriting and the use of inappropriate language hinder effective communication.
Be clear	Consider things before you speak or put anything in writing, to avoid misunderstandings or saying something you might later regret.
Avoid negativity	Take into account the other person's point of view. Effective communication is a two-way process and not a one-way street.
Get to the point	Don't communicate in a cold, cursory manner. The recipient should know that you care about them as individuals and are willing to take their concerns into account.
Listen	Don't wander off the subject or include unnecessary information. People are more likely to 'shut off', 'let their minds wander' or miss crucial points if too much information is given at any one time.

The building team

The recognised pattern by which the building team operates and communicates can be illustrated in a 'Hierarchy chart', which shows the level of authority and reporting lines in a construction project.

ACTIVITY

Below is a list of the main members of a building team. Using this, produce a chart to show the hierarchy of the members. The individual with most authority should be positioned at the top.

➤ suppliers (nominated and direct)
➤ principal building contractor
➤ subcontractors (nominated and domestic)
➤ health and safety inspector
➤ clerk of works
➤ local authority
➤ client
➤ quantity surveyor
➤ architect
➤ consulting (specialist) engineers

Receiving deliveries and general communication

No building site or construction contract could function effectively without a certain amount of day-to-day paperwork and form-filling, which enable the flow of information both within and between organisations.

ACTIVITY

You have supervised the delivery of materials shown on the note below. On checking the delivery, only 48 lengths of 50×50 were received, and several of the shrink-wrapped hardwood packages had splits in them.

BBS SUPPLIES
DELIVERY NOTE

Registered office
BRETT HOUSE
1 HAGELY ROAD
BIRMINGHAM
B11 N4

No. **8914**

Date **13 APRIL 2011**

Delivered to
T. JOYCEE
25 DAWNCRAFT WAY
STENSON
DERBY
D70

Invoice to
T. JOYCEE CONSTRUCTION
RIDGE HOUSE
NORTON ROAD
CHELTENHAM
GL59 1DB

Please receive in good condition the undermentioned goods

SAWN, TREATED SOFTWOOD
50 OFF   25 x 50 x 3600
50 OFF   50 x 50 x 4.800

KILN SEASONED HARDWOOD
25 OFF   25 x 150 x 2400 (REBATED WINDOW SILLS)

(SHRINK-WRAPPED IN PLASTIC)

Received by

Remarks

Note: Claims for shortages and damage will not be considered unless recorded on the sheet

1) Sign the delivery note and make any comments you think applicable.
2) Using a computer, design a letterhead for T. Joycee Construction and write a letter of complaint to the materials supplier concerning the short delivery and explain the potential implications of the damaged packaging.

Communicate workplace requirements efficiently

ACTIVITY

1) Why is clear and effective communication important to the team spirit and overall success of a company?

2) What essential information should be included in a telephone message for someone who was not available at the time of the call?

ACTIVITY

1) Explain the difference between nominated and domestic subcontractors.

2) State TWO advantages and TWO disadvantages of using visual, written or printed means of communication rather than verbal means.

3) Explain the difference between 'Time' and 'Day work' sheets.

Building information, quantities and communicating with others

ACTIVITY

Read the following statements and indicate whether they are true or false.

Statement	True	False
An assembly drawing shows detailed information concerning the arrangement of rooms in a building.		
Drawings and specifications are normally locked away in filing cabinets or plan chests in the site office, especially overnight and at weekends.		
Written descriptions of the material quality and standards of workmanship for a particular contract will be detailed in the standard conditions of the contract.		
A fully qualified carpenter is classified as a craft operative.		
The interrelationship between different tasks and the organisation of materials, labour and plant for a building contract can be obtained from the work programme bar chart.		
Written messages provide evidence that a communication has taken place, can be re-read if not understood the first time, can be passed on to others consistently, but can be lost or destroyed.		
The radius of a circle is equal to twice the diameter.		
'Archive' is the term given to both a collection of records and documents, as well as the place in which they are located.		
When ending letters, 'Yours sincerely' should be used, unless you have used the person's name in the greeting, in which case it should be 'Yours faithfully'.		
A 'variation order' is an instruction given by the architect to the principal contractor concerning any changes made to the original contract, including additions, omissions and alterations.		

Interpretation of information and general communication

 ACTIVITY

You are working on the construction of a house, where all except one of the windows to be used on-site is available from a standard range.

You require a non-standard size version of window CD OPP shown on the supplier's standard range sheet. The actual size required is 1250×1100 mm, for delivery to BBS construction on the Trent Valley Lakeside Housing Estate in Long Eaton, NG10 2BB.

Window type	Frame dimensions width x height	
V	634 x 921 634 x 1073 634 x 1226 634 x 1378	
V	921 x 921 921 x 1073 921 x 1226	Side hung
V	1221 x 1073 1221 x 1226	
CV AS CV OPP	1221 x 1073 1221 x 1226	
C AS C OPP	634 x 1073	
CD AS CD OPP	1221 x 1073	Fanlight opening
CVC	1808 x 1073 1808 x 1226	

Notes:
V = Ventlight   **C** = Casement   **OPP** = Opposite hand
D = Fixed light   **AS** = Hand as shown

Complete the following order form for the made-to-measure window.

ORDER FORM

FOR MADE TO MEASURE REPLACEMENT WINDOWS

When completing this form ensure the details show clearly on all copies.

MR/MRS/MS ....
                    (INITIALS)   (SURNAME)

DELIVERY ADDRESS ....

....

.... FULL POSTCODE ....

Please indicate where you would like the goods left, if delivered in your absence.

....

....

TELEPHONE: HOME   (STD ....) No ....

                       OFFICE (STD ....) No ....

I understand that the windows will be manufactured to the correct sizes based upon the dimensions I have provided and I accept responsibility for the dimensions

CUSTOMER'S SIGNATURE .... DATE ....

STORE USE ONLY

STORE   [0 | |]

DATE OF ORDER ....

PURCHASE ORDER No. ....

DRL No. ....

ADMIN. CHECKED BY ....

RECEIPT No. ....

TENDER TYPE: CHEQUE / CREDIT / CASH ....

Blue   – Order Copy
White   – Store Copy
Green   – Goods Inwards Copy
Yellow   – Customer Copy

SEE REVERSE SIDE FOR ORDERING INSTRUCTIONS AND GUIDANCE (Enter required details and complete all sections)

SKETCH YOUR WINDOWS SHOWING DESIGNS AND DIMENSIONS HERE. N.B. Always viewed from the outside

(See HOW TO ORDER WINDOWS notes for opening lights min/max)

(Please note that our range of Made to Measure Windows may vary in specification to our standard stock range.)

Do you require sills?   YES ☐   NO ☐

Do you intend to use these windows in conjunction with our range of standard windows?   YES ☐   NO ☐

Have you ordered made to measure windows from us before? ☐ If so, please state approx. date of order ....

QUICK QUIZ

1. What does the abbreviation 'bwk' on a drawing refer to?
   a. building
   b. boarding
   c. binding
   d. brickwork

2. Symbols are used on building drawings to:
   a. make them easier to understand.
   b. indicate the company who designed the building.
   c. reduce the need to include technical information.
   d. represent different material and components.

3. What would 20 mm on a 1:10 scaling drawing represent?
   a. 20 mm
   b. 100 mm
   c. 200 mm
   d. 2000 mm

4. General location plans are often drawn using scales of:
   a. 1:100, 1:50 or 1:20
   b. 1:125, 1:1250 or 1:2500
   c. 1:50, 1:100 or 1: 200
   d. 1:1, 1:5 or 1:10

5. Details of all doors and associated ironmongery required for a particular job would be shown on a:
   a. specification.
   b. variation order.
   c. schedule.
   d. confirmation notice.

6. Which one of the following documents provides a measure of the quantities of labour and material required at the tendering stage?
   a. specification
   b. conditions of contract
   c. schedule
   d. bill of quantities

7. Which one of the following is not a preferred measurement of length in the construction industry?
   a. millimetre
   b. centimetre
   c. metre
   d. kilometre

8. The correct formula for calculating the perimeter of a rectangle is:
   a. 2(length + breadth).
   b. 2(length × breadth).
   c. length + breadth.
   d. length × breadth.

9. If the length of a room is 6500 mm and the total perimeter measures 20 m, what is the width of the room?
   a. 6.5 m
   b. 3250 mm
   c. 7000 mm
   d. 3.5 m

10. Which one of the following is an advantage of using verbal communication?
    a. It may not be the most effective where large numbers of people are involved.
    b. It enables a quick response and gives direct personal contact.
    c. It can be passed on inconsistently or forgotten altogether.
    d. Disputes can arise at a later stage, as there are no written records that can be referred to.

UNIT 1003

Building methods and construction technology

All buildings are constructed using similar basic principles and contain certain common **elements**. Whatever the type of building, they will all have a substructure and superstructure consisting of suitable foundations, walls, floors and a roof.

Key knowledge:
➤ foundations, walls and floor construction.
➤ construction of internal and external masonry.
➤ roof construction.

key terms

Elements: the constructional parts of a building's substructure and superstructure. They are classified into three main groups: primary elements, secondary elements and finishing elements, depending on their function.

Building: the act or process of creating and maintaining the built environment.

Construction: the process of building or assembling the built environment.

Buildings: structures that enclose space and in doing so create an internal environment. The actual structure of a building is termed the external envelope, which protects the internal environment from the outside elements, known as the external environment.

Foundations: the part of a structure that transfers the loads of the superstructure safely onto the supporting ground.

Walls: the vertical enclosing and dividing elements of a building.

Floors: The horizontal ground and upper levels in a building that provide a walking surface.

Roof: the uppermost part of a building, that spans the walls and protects the building and its contents from the effect of weather.

Foundations, walls and floor construction

Bench and datum marks

ACTIVITY

Bench and datum marks are identified points from which all other positions on-site or in buildings are either taken or are related to.

The illustrations show bench and datum marks. Name each and explain where they may be found or used.

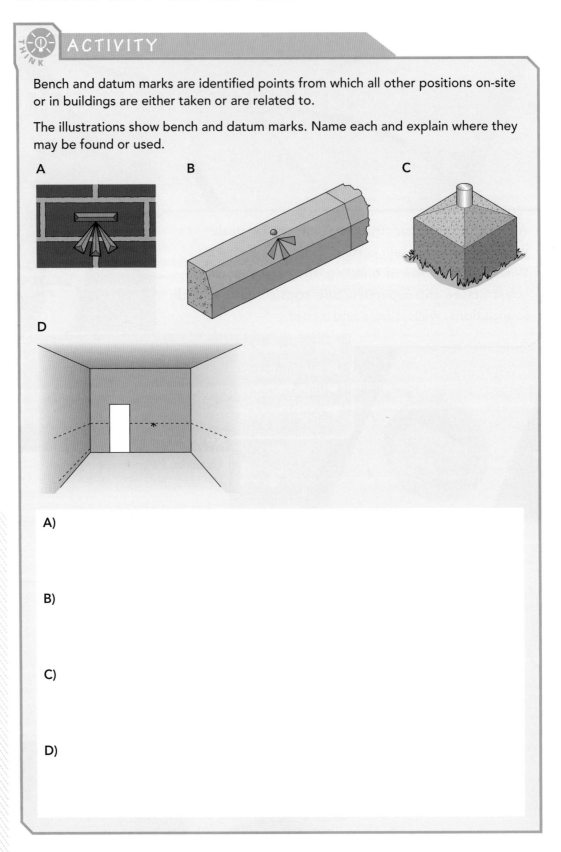

A

B

C

D

A)

B)

C)

D)

Types of foundations

Foundations are the part of the **substructure** that transfers the loads of the **superstructure** safely down onto a suitable load-bearing layer of ground.

ACTIVITY

Name the different types of foundations shown in the illustrations and state a typical use.

A) Name:

   Use:

B) Name:

   Use:

C) Name:

   Use:

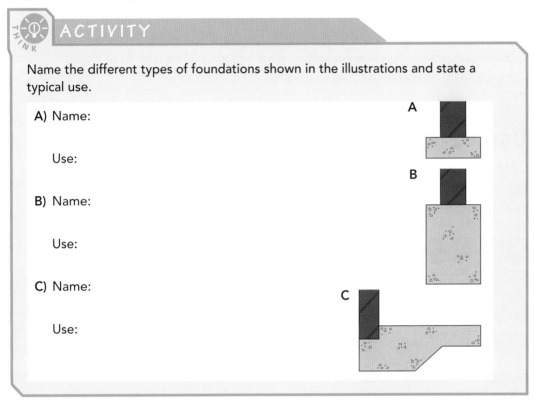

key terms

Substructure: the structure from below the ground up to and including the ground floor. Its purpose is to receive the loads from the main building superstructure and its contents, and transfer them safely down to a suitable load-bearing layer of ground.
Superstructure: all of the structure above the substructure, both internally and externally. Its purpose is to enclose and divide space, and transfer loads safely to the substructure.

Materials used in concrete

Concrete (a mixture of cement, fine aggregate, coarse aggregate and water) is the main material used to form foundations and floors.

ACTIVITY

1) Describe the difference between fine and coarse aggregates.

2) Explain what each of the main materials used in concrete does when mixed.

3) State the reason steel reinforcement has been added to the strip foundation shown.

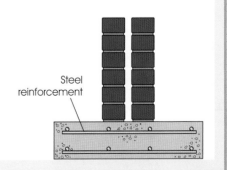

Steel reinforcement

I've heard that the water used in concrete should be of a 'potable' quality. What is potable and why should it be used?

Potable refers to a liquid that is suitable for drinking because it's clean and uncontaminated. The water used for concrete should be free from any impurities that could affect its strength.

Use of DPMs and DPCs

DPMs and DPCs are waterproof barriers that are incorporated into walls and solid ground floors to prevent dampness coming up from the ground and penetrating into the building.

 ACTIVITY

1) State what the abbreviations DPM and DPC stand for.

   DPC

   DPM

2) State the main difference between a DPC and a DPM.

3) Name a common material used for both DPMs and DPCs.

Foundations, walls and floor construction

ACTIVITY

1) Describe and state the purpose of a building's substructure.

2) Explain what an OBM is and how it relates to a site datum.

3) Explain why DPCs are positioned at least 150 mm above the exterior ground level.

4) State the circumstances in which a raft foundation may be used instead of a strip foundation.

5) State the proportions of a typical concrete mix for unreinforced strip foundations and ground-floor slabs.

Construction of internal and external masonry

Types of wall

The walls of a building may be classed as either load-bearing or non-load-bearing. External walls have an enclosing role; internal walls have a dividing one and are normally termed as partitions. Openings in load-bearing walls for windows and doors are bridged by either arches or lintels, which support the weight of the wall above.

REFLECT ACTIVITY

Look at the illustrations and answer the following questions.

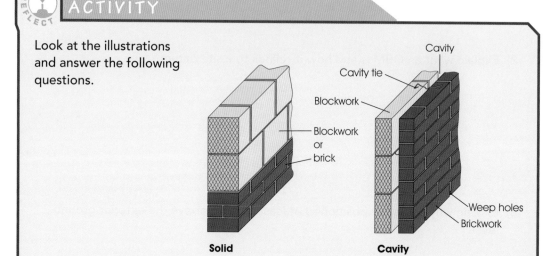

Solid

Cavity

1) State why solid external walls are not commonly used in new buildings.

2) Explain what must be done to the cavity if it extends down to the foundation.

3) How is moisture prevented from travelling across the cavity ties?

4) What is the purpose of weep holes?

Bonding

The strength of brickwork and blockwork is dependent on its bonding, which is the overlapping of vertical joints to spread any loading evenly throughout the wall. The actual overlap or bonding pattern will vary depending on the type of wall and the decorative effect required.

ACTIVITY

Produce sketches in the space below to show the difference between English, Flemish and stretcher bonds used in brickwork.

English:

Flemish:

Stretcher:

Internal walls

Internal walls are constructed either as solid or framed walls.

➤ Solid internal walls consist of bonded blocks or bricks bedded in cement mortar.

➤ Framed internal walls are built using timber or metal studs fixed between sole and head plates, often incorporating noggins for stiffening and to provide fixing points.

 ACTIVITY

The illustrations show internal walls grouped according to their method of construction or finish. Name each group and state a situation where they might be used.

	Group	Situation
Finish plaster, Plasterboard, Dabs of adhesive, Undercoat plaster		
Noggin, Stud, Sole, Plasterboard nailed to timber partition, Plasterboard screwed to metal		

Your questions answered...

Why do the concrete lintels on my site have a 'T' marked on one face?

Concrete lintels contain steel reinforcement towards their bottom edge in order to resist tensile forces. The 'T' indicates the top of the lintel, and it is essential that they are handled and built in with this uppermost, otherwise they may simply fold in two.

Construction of internal and external masonry

 ACTIVITY

1) Explain the main reason cavities are included in external walls.

2) State the normal thickness of the brickwork outer leaf of a cavity wall.

 ACTIVITY

1) Name TWO materials that are used for lintels.

A)

B)

2) What is a suitable mix ratio for brickwork mortar? Explain how the parts should be measured.

Roof construction

Roof shapes

Roofs are classified according to the slope or pitch of the roof surface, by their shape and also by their method of construction.

ACTIVITY

Name the roof shapes shown in the illustrations.

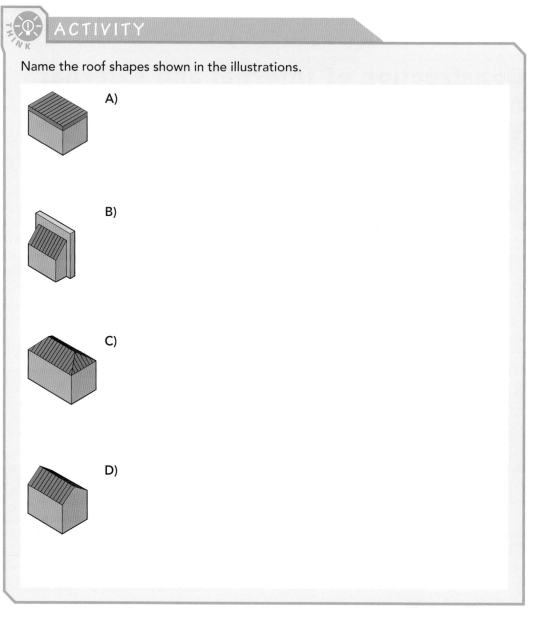

A)

B)

C)

D)

Roof structures and components

Timber pitched roofs may be divided into two main types, depending on their method of manufacture.

➤ Modern pitched roof structures use triangular roof frames called trussed rafters, which are prefabricated off-site under factory conditions and delivered to the site ready for erection.

➤ Traditional framed cut roofs are almost entirely constructed by carpenters on-site from loose timber sections, utilising simple jointing methods.

 ACTIVITY

Draw a line between the appropriate boxes to connect the roof components with the correct description.

Spreader plates	The main load-bearing timbers in a roof, which are cut to fit the ridge, and bird's-mouthed over the wall plate.
Flashing	The backbone of the roof, which provides a fixing point for the tops of the rafters, keeping them in line.
Jack rafters	Transfers the loads imposed on the roof uniformly over the supporting brickwork and also provides a bearing and fixing point for the feet of the rafters and ceiling joists.
Hip rafters	Used where two sloping roof surfaces meet at an external angle, to provide a fixing point for the jack rafters and transfer their loads to the wall plate.
Common rafters	Span from the wall plate to the hip rafter, like common rafters that have had their tops shortened.
Ridge board	A triangular roof frame consisting of rafters, ceiling joist and structural triangulation, fastened with galvanised metal nail plates.
Wall plates	The rafter that is used in the centre of a hip end.
Crown rafters	Span from the ridge to the valley, like common rafters that have had their feet shortened (the reverse of jack rafters).
Purlin	Used at an internal angle to provide a fixing point for the cripple rafters and transfer their loads to the wall plate.
Struts	A beam that provides support for the rafters in their mid-span.
Binders and hangers	Two rafters with noggins nailed between them, which are fixed to the last common rafters to form the overhanging verge on a gable roof.
Cripple rafters	Act as ties for each pair of rafters at wall plate level, as well as providing a surface on which the ceiling plasterboard can be fixed.
Ceiling joists	Used to stiffen and support the ceiling joists in their mid-span, preventing them from sagging and distorting the ceiling under load.
Valley rafters	Used to stiffen and support the ceiling joists in their mid-span, preventing them from sagging and distorting the ceiling under load.
Ladder frames	Provide a suitable bearing for the struts at ceiling level.
Collar ties	Used to prevent the spread of rafters in the same way as ceiling joists, but at a higher level and may also be used to provide some support for the purlins.
Trussed rafter	The pieces of sheet metal, normally lead or zinc, which are attached around the joints and angles of a roof to protect against rainwater penetration.

Metal fixings used in roof construction

A range of metal fixing devices are used in roof construction, either to join component parts together or provide a positive tie between the roof and walls at the eaves and verge.

ACTIVITY

Name each of the fixings shown in the illustrations and state a typical use.

A)

B)

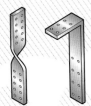

C)

D)

Roof construction

ACTIVITY

1) State the main difference between flat and pitched roofs.

2) Describe what is meant by the 'eaves' of a roof.

3) Explain the difference between traditional and modern pitched roofs.

ACTIVITY

State TWO reasons why wall plates are used in roofs.

1)

2)

ACTIVITY

State TWO ways of forming the slope on flat roofs.

1)

2)

Your questions answered...

What is sarking felt and where is it used?

Sarking felt is the felt laid over the rafters of pitched roofs before the tile battens are fixed, to prevent moisture entering the building in the event that the covering tiles or slates get damaged or joints bypassed by wind-driven rain.

Produce sketches to show typical building details

 ACTIVITY

Using the following specifications, make a sketch below to show a cross-section through a storey building and label all the component parts.

➤ **Foundations:** narrow concrete strip.

➤ **External walls:** cavity wall, comprising brick outer leaf and blockwork inner leaf, cavity ties and DPC.

➤ **Ground floor:** solid construction, comprising hardcore, blinding, DPM, concrete over-site, insulation and cement screed.

➤ **Roof:** pitched construction with overhang eaves, comprising common rafters, ceiling joists, binder, hangers, wall plates, tie-down straps, tiling battens on sarking felt, concrete roof tiles and ridge capping.

Building methods and construction technology word search

ACTIVITY

Solve the following clues and find these words hidden in the word square.

1) A brick showing its longest face. (9)

2) A brick showing its shortest face. (6)

3) A vertical joint in brickwork. (6)

4) A horizontal joint in brickwork. (3)

5) Material used in walls to prevent the rise of moisture. (4,5,6)

6) An item used in walls to bridge openings. (6)

7) A wall used internally to divide space. (9)

8) The process of applying plasterboard to a wall surface with dabs of adhesive. (3,6)

9) Large aggregate used in concrete. (6)

10) Quality of water used in concrete and mortar mixes. (7)

11) Foundation that covers the footprint of a building. (4)

12) A component used to join the two leaves of a cavity wall. (3)

13) Material used to bond bricks and blocks. (6)

14) A bond consisting of alternate rows of bricks laid lengthways and bricks laid width-ways. (7)

15) The process of finishing mortar joints. (8)

16) The measuring of materials used in a mortar to achieve the required strength and specification. (7)

17) The vertical member in a timber or metal partition wall. (4)

18) Board used to finish the verge of a pitched roof. (5)

19) A roofing component that provides the fall on a timber-joisted flat roof. (7)

20) The beam used in pitched roofs to provide intermediate support to rafters. (6)

G	W	C	H	D	C	W	S	T	U	D	F	W	F	Q	R	Q	R
N	X	J	S	H	N	N	T	B	W	A	V	O	I	B	V	G	W
N	V	Z	I	I	Q	T	R	K	L	B	Q	D	E	B	M	C	F
Y	T	P	L	Y	F	W	E	T	T	M	N	A	F	I	R	F	W
Q	G	R	G	G	F	S	T	J	M	E	F	M	O	K	F	S	Z
A	U	P	N	B	T	F	C	F	P	N	N	P	I	J	B	B	K
P	B	R	E	D	A	E	H	R	G	N	E	P	P	P	A	G	L
Z	Y	L	I	R	G	H	E	V	B	X	V	R	A	T	R	O	M
W	H	U	T	Y	R	P	R	S	S	N	E	O	T	Y	G	A	V
S	N	J	F	L	D	A	W	A	R	W	B	O	T	R	E	Q	T
X	U	B	V	I	F	K	R	G	F	G	E	F	R	Q	D	A	G
E	P	O	I	N	T	I	N	G	H	T	B	C	R	W	C	D	A
D	J	G	E	I	S	R	E	L	B	A	T	O	P	R	U	I	U
C	M	D	T	N	H	S	F	I	B	D	X	U	Q	F	S	S	G
R	I	X	B	G	R	S	T	N	G	N	I	R	R	I	F	U	I
F	K	A	P	A	R	T	I	T	I	O	N	S	A	S	C	M	N
V	O	W	O	J	Y	E	T	E	I	G	K	E	C	R	R	E	G
T	L	C	N	W	E	H	E	L	G	O	L	G	H	T	W	T	Q

QUICK QUIZ HOW MUCH DO YOU KNOW ABOUT BUILDING METHODS AND CONSTRUCTION TECHNOLOGY.

1. What is the main difference between bricklaying mortar and concrete?
   a. The amount of water used.
   b. The type of aggregate used.
   c. The type of cement used.
   d. There is no difference.

2. Bench and datum marks are where:
   a. walls of a building intersect.
   b. buildings are set out from.
   c. levels are taken from.
   d. foundations are required.

3. The most common type of foundation used for low-rise domestic structures is:
   a. rafts.
   b. narrow strips.
   c. wide strips.
   d. trench fill or deep strips.

4. What is the main difference between DPCs and DPMs?
   a. The type of material used.
   b. The size of material used.
   c. Their purpose.
   d. Their location.

5. The main purpose of bonding brickwork is to:
   a. provide a surface that plaster will bond to.
   b. spread the loading evenly throughout the wall.
   c. provide a decorative finished effect.
   d. minimise the use of materials.

6. Common rafters in a pitched roof are fixed at either end to:
   a. sole plate and purlin.
   b. wall plate and ridge.
   c. soleplate and crown.
   d. wall plate and soffit board.

7. What type of bonding is used in cavity walls?
   a. Cavity bond
   b. English bond
   c. Stretcher bond
   d. Flemish bond

8. The component that is used to bridge openings in walls is a:
   a. lintel.
   b. joist.
   c. binder.
   d. sleeper.

9. Which ONE of the following statements is false?
   a. Brickwork can be pointed with a stronger mortar to increase weather resistance in exposed locations.
   b. Gauging is the measuring of materials used in a mortar to achieve the required strength and specification.
   c. Cavity wall ties include a design feature, such as a twist or bend in the centre that prevents moisture passing between the two leaves of the wall.
   d. Sharp sand is normally used for brickwork mortar mixes.

10. Lateral restraint straps are used in roofing to:
    a. fasten the joints of trussed rafters.
    b. provide a positive tie between roof and walls.
    c. provide a positive tie between rafters and wall plates.
    d. stiffen and support the ceiling joists in their mid-span.

UNIT 1004

Produce basic woodworking joints

Joint making is regarded as a measure of a woodworker's skill, since it requires the mastering of a variety of very accurate marking and cutting techniques. In training, joint making instils a 'feel' for both materials and the use of hand tools, which will not be lost in later years, even when progressing on to machines and powered hand tools.

Key knowledge:
- ➤ selecting and using hand tools to produce basic woodworking joints.
- ➤ selecting and storing materials used to produce basic woodworking joints.
- ➤ marking out to produce basic woodworking joints.
- ➤ forming basic woodworking joints.

Select and use hand tools

Bench equipment

Woodworkers use a range of equipment to hold and support the work piece or tool while preparing timber and cutting joints. Name each item shown in the illustrations and state what it is used for.

A)

B)

C)

D)

A

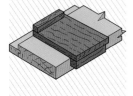

B

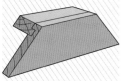

C

D

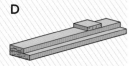

Selecting hand tools

ACTIVITY

Before undertaking any work task, you should ensure you have the necessary tools available to hand.

Produce a list of TEN tools you would require to mark out and produce from sawn timber a rectangular frame with mortise-and-tenon joints.

1)

2)

3)

4)

5)

6)

7)

8)

9)

10)

Select and use hand tools

 ACTIVITY

1) Describe how the handsaw is positioned and given a start when undertaking ripping operations.

2) Name a type of chisel that is intended for handwork only and should never be driven using a mallet or hammer.

3) State the possible reason why the corners of a smoothing plane blade are digging into the timber surface, leaving plane marks while cleaning up.

 ACTIVITY

State TWO signs that would indicate that a saw is dull (blunt).

1)   2)

 ACTIVITY

What size of mortise chisel should be selected to mortise the door stile illustrated?

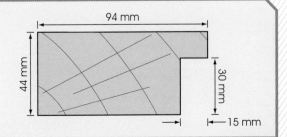

Timber conversion

Conversion is the sawing up or breaking down of the tree trunk into various-sized pieces of timber for a specific purpose.

ACTIVITY

The illustrations show the four main ways used to convert tree trunks. Name each method and describe the type of timber produced.

A)

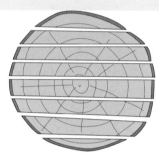

B)

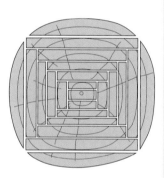

C)

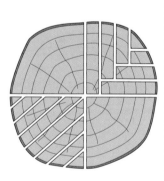

D)

Identification and properties of timber

Timber can be classified as either a 'hardwood' or a 'softwood'. These terms can be misleading as they are not based on the hardness or strength of the timber, but on botanical differences in the tree's cellular structure.

ACTIVITY

Look up the different types of wood, using the internet if you wish, and complete the table, stating whether it is hardwood or softwood and listing some common uses.

Wood	Classification	Common uses
Douglas Fir		
Redwood		
White wood		
Ash		
Oak		
Mahogany		

Timber seasoning

Seasoning refers to the controlled drying by natural or artificial means of converted timber sections, to a moisture content that is similar to the surroundings in which it will be used.

ACTIVITY

Complete the table to show a suitable moisture content for each location.

Location	Moisture content
Carcassing timber such as joists and rafters	
External joinery	
Internal timber in an occasionally heated building	
Internal timber in a continuously heated building	
Internal timber positioned over or near a source of heat	

Timber-based board material

Timber-based board materials are specified in a variety of situations, such as floor and roof coverings, wall panelling, panels in framed joinery and the construction of units and fitments. They are available in large sheet sizes and a range of widths, grades and surface finishes to suit the work in hand.

ACTIVITY

1) Produce sketches to show the difference between the following manufactured boards:

Plywood (three-ply)	Plywood (Stout-heart)
Plywood (Multi-ply)	**Block board**

Your questions answered...

I thought the standard size of timber-based board material was 2440 × 1220 mm. However, I've seen a specification for a job that states that 1220 × 2440 mm sheets of veneered MDF are to be used. What does this mean?

It is standard practice when stating the sheet size of plywood and other veneered boards that the grain direction runs parallel to the first stated dimension. Thus a 2440 × 1220 mm board will have long grain and a 1220 × 2440 mm board will have short grain.

ACTIVITY

2) What common names are used for particle boards and fibreboards? Describe the main differences between them.

Storing materials correctly

ACTIVITY

In a small group, consider the following real-life scenario.

Karen and Jade work for a small joinery company that mainly manufactures and installs joinery fitments. They have been given the job of making a radiator cover using kiln-seasoned hardwood for the main framework. However, the only timber available has been stored haphazardly in an external rack. On selecting the timber Jade notices that some of it feels damp and most is also distorted.

Karen suggests that it should be okay to use the drier pieces, and that they could cut up the lengths to remove most of the distortions.

➤ What do you think of Karen's suggestion?
➤ What are the potential implications of using this timber?
➤ What storage procedures should be followed in the future to prevent a re-occurrence of the situation?

ACTIVITY

State THREE of the main differences between softwood and hardwood.

1)

2)

3)

ACTIVITY

1) Explain what the terms 'wrot' and 'un-wrot' mean.

2) State the purpose of using piling sticks or cross bearers when stacking timber.

3) Explain why sheet material and items of joinery should not be stored leaning against a wall.

4) Describe the main differences between 'air' and 'kiln' seasoning.

Marking out

Your questions answered...

What is the difference between setting out and marking out, and who does what?

Setting out involves looking at the architect's plans and assembly drawings, specifications and their own survey details and translating them into full-size workshop rods showing the vertical and horizontal sections of the item.

Marking out involves referring to design drawings, workshop rods and cutting lists produced during the setting-out process and the selection and marking out of timbers to show the exact position of joints, mouldings, sections and shapes for a particular item.

Who does what will depend on the size of joinery works and the volume of work it handles. Setting and marking out may be undertaken by one person or treated as separate roles to be undertaken by different people. In some organisations the setting and marking out and assembly of a joinery item is undertaken by the bench joiners.

Marking out tools

Timber cannot be accurately marked out for cutting until it has been first prepared with flat faces and square parallel edges and face and edge marks have been applied to indicate that the two adjacent surfaces are true and square to each other.

The illustrations show a range of tools that can be used when marking out. Name each tool and state its typical uses.

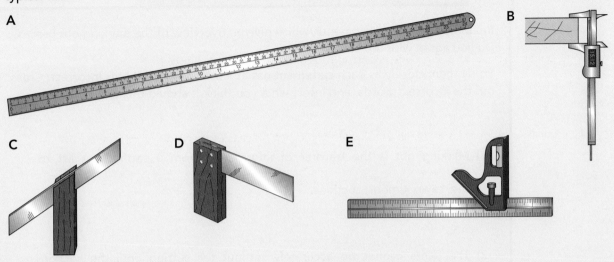

A

B

C

D

E

	Name of tool	Typical uses
A		
B		
C		
D		
E		

Do you know what marking out involves?

Read the following statements, which give an overview of the marking out process. Do you agree with the statements?

Study them closely as each statement has at least one word that is incorrect. Cross out the incorrect words, and insert what you think it should be above them

1) Marking out is the transfer of information from a setting out list to the sawn timber sections.

2) The joints cannot be accurately set out for cutting until the sections have been first prepared with flat faces and square parallel edges.

3) Face and edge marks are applied to indicate that the two opposite surfaces are in-wind and square to each other.

4) Before marking out, the rod and assembly drawings should be checked for accuracy and consistency. Any irregularities found don't need resolving before proceeding with the work.

5) Wrong information transferred on to the rod at this stage will result in errors later during setting out, which are likely to be time-consuming and very inexpensive.

Use of a try square

Try squares are used to test pieces of wood, made-up joints and so on for square, as well as being used when marking lines at right angles across faces or edges of a piece of wood. To ensure their accuracy they should be periodically checked for trueness.

ACTIVITY

Produce a sketch and brief notes to show how a try square can be tested for accuracy.

Marking out

ACTIVITY

1) Explain the importance of face and edge marks that are applied to a timber section before marking out joints.

2) State the reason shoulder joints may be marked out using a marking knife.

3) State TWO things to be considered when setting up a gauge to mark the thickness of a tenon.

4) State why workshop rods and assembly drawings should be checked before marking out is started.

5) Name the tool or piece of equipment that is best used to square lines around pre-moulded timber sections.

Woodworking joints

All woodworkers make joints. The carpenter makes joints that are normally load-bearing. The joiner uses mainly framing joints for doors, windows and decorative trims. The cabinet/furniture maker also uses framing joints, both to create flat frames and also to build up three-dimensional carcasses. In addition, both the joiner and the cabinet/furniture maker may be involved with a range of jointing methods for manufactured boards.

ACTIVITY

The illustrations show a range of joints. Complete the table below, naming the joints and stating a typical use for each.

	Name of joint	Typical uses
A		
B		
C		
D		
E		

A

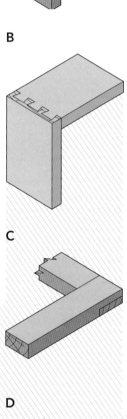

B

C

D

E

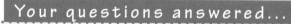

Your questions answered...

What does 'dry assembly' mean and why should I do it?

Dry assembly is the term used to describe the assembly of joinery without the use of adhesive. It's considered good practice to dry assemble all joinery items during manufacture, as it provides the opportunity to check the fit of joints, overall size, square and winding.

Timber defects

Defects are faults that are present in timber. Some cause structural weakness and others may spoil its appearance. These defects can be divided into two groups: timber processing/seasoning defects and natural defects. As far as possible, their cause should have been avoided or the result cut out during its conversion.

ACTIVITY

Draw a line between the appropriate boxes to connect the image of a defect with the correct name.

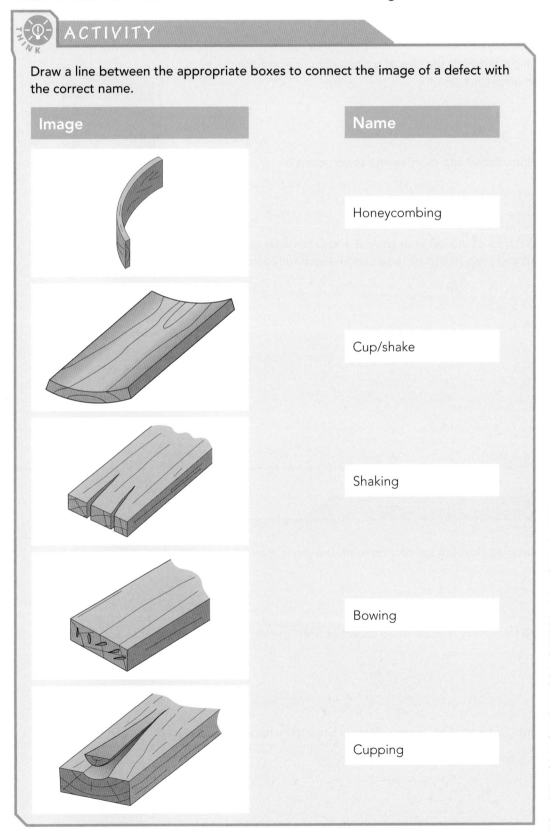

Image	Name
	Honeycombing
	Cup/shake
	Shaking
	Bowing
	Cupping

Joint selection and proportions

For strength reasons the proportions used when marking out and cutting joints should be related to the sectional size of the members being joined.

The illustration shows the elevation of a framed panel door.

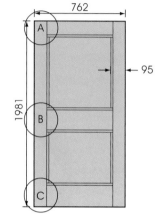

1) What is the relationship between the size of a timber section and the thickness and width used for a tenon?

2) Why are haunches introduced into mortise and tenon joints?

3) Produce labelled sketches of the rail ends at A, B and C to name each joint and show the layout and proportions of tenon and haunch for each.

   A)   B)   C)

1) State why, when making a dovetail tee halving joint, the pin is marked and cut before the socket.

2) 'PVA' adhesive used for joinery may be rated as either 'INT' or 'MR'. State what all three abbreviations stand for.

3) Explain what the term 'rubbed joint' means. State a situation where it may be used.

Basic woodworking joints

ACTIVITY

Solve the following clues and find these words hidden in the word square. You may find the words written forwards, backwards, up, down or diagonally.

1) The process of drying timber. (9)
2) A tool commonly used to mark square lines around a timber section. (3,6)
3) The part of the joint commonly used in the head of a window frame. (7)
4) An item of equipment used when sawing wood on a bench. (5,4)
5) A joint commonly used for door linings. (7)
6) Freshly felled unseasoned is often termed as... (5)
7) Movement in timber when its moisture content is reduced. (9)
8) The sawing up of a tree trunk. (10)
9) Best method of sawing timber for joinery use. (7)
10) A drawing showing how an item is put together. (8)
11) A plane used to smooth and clean up sawn curves. (7)

12) A long bevelled edge chisel for handwork only. (6)
13) A finishing tool used to clean up irregular grain. (7)
14) The name for defects such as knots, cup shakes and star shakes. (7)
15) A saw used to cut the shoulders of joints. (5)
16) The middle word in the abbreviation in MDF. (7)
17) A curvature across the width of a tangentially sawn board. (7)
18) A manufactured board material made from pulped wood. (10)
19) The marking out tool used to score cutting lines parallel to the edge of a piece of timber. (5)
20) A dovetail joint used for drawer fronts. (6)

H	S	D	I	E	N	W	S	T	U	D	T	W	F	Q	D	Q	R
N	X	W	S	H	E	W	O	R	A	R	V	O	T	B	V	G	W
S	V	Z	I	I	Q	T	R	K	Y	B	Q	D	E	E	M	C	F
Y	T	P	M	O	R	T	I	S	E	M	N	C	F	G	N	N	W
G	G	R	G	F	F	S	Q	E	M	E	F	O	O	K	F	O	Z
K	U	P	N	I	T	U	C	A	S	S	E	M	B	L	Y	I	N
P	B	R	E	B	A	E	H	S	G	N	E	P	P	P	A	S	R
Z	Y	L	I	R	G	G	E	O	B	X	V	A	A	T	R	R	M
D	K	U	E	E	R	P	N	N	S	N	E	S	T	R	G	E	V
S	O	M	F	B	D	A	W	I	R	W	B	S	S	R	I	V	Y
R	O	A	V	O	F	S	R	N	P	G	E	F	H	Q	D	N	S
E	H	O	I	A	C	I	N	G	H	P	B	O	R	W	C	O	G
D	H	G	E	R	S	R	E	R	B	A	U	O	I	R	U	C	A
L	C	D	A	D	H	S	F	E	B	S	X	C	N	F	A	S	U
R	N	P	B	G	R	S	T	E	I	N	I	R	K	I	F	U	G
F	E	L	A	R	U	T	A	N	R	E	T	R	A	U	Q	M	E
R	B	W	O	H	Y	E	G	E	I	G	K	E	G	R	R	E	N
E	L	B	N	W	E	Y	T	I	S	N	E	D	E	P	P	A	L

1. The most suitable joint for lengthening a wall plate is a:
   a. heading joint.
   b. housing joint.
   c. mortise-and-tenon joint.
   d. dovetail joint.

2. Knots, star shakes and reaction wood are known as:
   a. seasoning defects.
   b. manufacturing defects.
   c. construction defects.
   d. natural defects.

3. The best method of conversion for joinery purposes is:
   a. quarter sawn.
   b. tangential sawn.
   c. rip sawn.
   d. through and through sawn.

4. In the absence of a paring chisel, which of the following would be most suitable for paring timber?
   a. firmer
   b. mortise
   c. bevelled edge
   d. registered

5. Which one of the following is most suitable for cleaning up joints after assembly?
   a. block plane
   b. smoothing plane
   c. jack plane
   d. shoulder plane

6. The part of a tree that produces the most usable timber is the:
   a. canopy.
   b. branches.
   c. trunk.
   d. roots.

7. Which one of the following statements is false?
   a. Kiln drying is a method of seasoning timber.
   b. Window frames can be jointed using a mortise and tenon.
   c. A block plane is good for cleaning up end grain.
   d. Freshly felled unseasoned timber is often termed 'blue timber'.

8. Which one of the following is most suitable to use in conjunction with a bench hook for cross-cutting joint shoulders?
   a. panel saw
   b. tenon saw
   c. cross-cut saw
   d. coping saw

9. Scarf joints are mainly used to:
   a. lengthen structural timber.
   b. join boards in width.
   c. frame up joinery items.
   d. edge sheet material.

10. Timber manufactured boards such as chipboard, MDF and plywood are best stored:
    a. on edge leaning against a wall.
    b. flat on a level floor.
    c. flat on timber cross bearers.
    d. on end leaning against a wall.

UNIT 1005

Maintain and use carpentry and joinery hand tools

The skills required to use and maintain hand tools are relatively easy to acquire. However, they can take much patience and practice to perfect, but the rewards will be with you for life. The tool skills that you develop, coupled with good woodworking practices, will be evident in the quality of your finished product, for better or worse!

Key knowledge:
➤ maintaining and storage of hand tools.
➤ using handsaws.
➤ using hand-held planes.
➤ using hand-held drills.
➤ using woodworking chisels.

Tool maintenance

Woodworkers use a range of hand tools to mark out, saw, plane, bore, shape and hold timber. They must follow the correct **maintenance** and storage procedures, which will ensure long and efficient service.

Handsaws

ACTIVITY

Fill in the gaps in the following sentences.

1) In order to prevent a saw blade from _____ in the timber,

   its _____ are _____, that is each alternate tooth is

   _____ outwards to make a saw cut, or _____,

   just wide enough to clear the blade.

2) Teeth for _____ are _____ end teeth and the cutting

   action is, in fact, like a series of tiny chisels, each cutting one behind the other.

3) Teeth for _____ have _____ to sever the fibres of the

   timber, which are so arranged that they cut two knife lines close together.

4) To make the saw cut smoothly, the front faces of the teeth slope forward. This is termed as the

   _____ and varies from 3° for _____ saws and up to 14°

   for _____ saws.

Blades

ACTIVITY

1) The illustration shows the two angles that are used when grinding
   and honing plane and chisel blades. Name and state a suitable
   angle for each (in degrees).

   A)

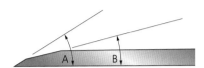

   B)

2) The shape of a blade's cutting edge will depend on its use: most will be square but some are not.
   Produce sketches to show TWO other shapes that may be used and explain the reasons why.

Your questions answered...

What is the best sharpening stone to use for plane and chisel
blades?

It's really down to your budget and personal preference. Oilstones are
normally less expensive, but waterstones and diamond stones are cleaner to
work with as they are both lubricated with water.

Quick check: Maintain and store hand tools

1) Name THREE types of stone used for honing plane and chisel blades, stating for each a suitable lubricant.

2) Explain in sequence the grinding and honing of a firmer edge chisel.

3) Either oil or water is used on sharpening stones:
   A) state the reason why

   B) state an advantage and disadvantage of each

   C) name the type of stone each is suitable for.

4) State how saws should be stored to prevent damage to teeth, distortion and rusting.

5) State why a plane or chisel blade may be further honed on a thick piece of leather.

Use carpentry and joinery handsaws

Types of handsaw

Handsaws are classified by the shape of their teeth, the length of the cutting edge and also the number of teeth per 25 mm.

ACTIVITY

A

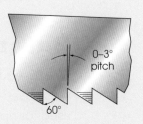

0–3°
pitch

60°

B

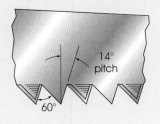

14°
pitch

60°

1) Look at the two hand saws that are illustrated above, and state what each is best used for.

A)

B)

2) The table lists handsaws according to their blade length and number of teeth per 25 mm. Complete the table to name the type of saw, and state what it is typically used for.

Name of saw	Teeth per 25 mm	Blade length	Typical use
	3–6	Up to 750 mm	
	6–8	550 to 650 mm	
	8–10	500 to 550 mm	

I have been given a handsaw that has a slightly curved blade and its teeth extend around the radius on its front end. What is its name and intended use?

It's called a floorboard saw and is used for cutting heading joints when forming access points in existing timber-boarded floors.

Cutting action

Saws used for ripping timber along the grain require a different cutting action to those used for cross-cutting. At a push, saws intended for cross-cutting can be used for ripping, albeit slowly and inefficiently, but it's almost impossible to use a ripsaw for cross-cutting as it will 'jar' and tear the wood.

 ACTIVITY

Produce sketches to show the difference in the cutting action and shape of teeth used for cross-cutting and ripping.

Carpentry and joinery handsaws

 ACTIVITY

1) State the reason backsaws have a folded metal strip on their upper edge.

2) State TWO typical uses for both padsaws and coping saws.

3) Explain the advantages of having a 'Hardpoint' saw in your tool kit.

4) State what can be done to ease the rip-sawing of case-hardened timber that starts to bind on the saw blade.

Use carpentry and joinery hand-held planes

Hand-held planes

Woodworkers use hand planes to remove paper-thin shavings. Some are used to produce flat, straight and true surfaces, while others are used to produce rebates, grooves, moulding curves and for cleaning up.

ACTIVITY

The illustrations show the three types of bench plane in common use. Name and state a typical use for each plane.

A)

B)

C)

A

B

C

The illustration shows an exploded view of a typical bench plane.

Complete the list to name each numbered part.

1)

2)

3)

4)

5)

6)

7)

8)

9)

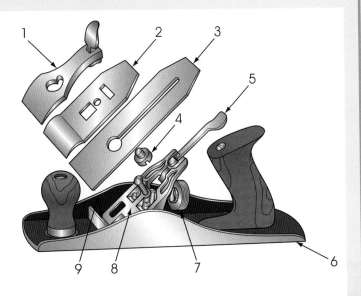

Explain what has gone wrong while rebating the section of timber shown in the illustration.

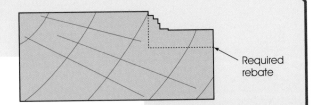

Required rebate

Specialist planes

With the exception of the compass plane, most other specialist planes have their cutting irons set with their bevel side up and no back iron. The bevel on the cutting iron acts in place of a back iron to break the wood shaving.

  ACTIVITY

Draw a line between the appropriate boxes to connect the names of specialist planes with the appropriate description or use.

Router plane	Cleaning up and truing rebates and the end grain shoulders of tenon joints
Compass plane	Has a flexible sole, enabling it to be adjusted to fit against internal and external curved shapes
Spokeshave	Used to ease or clean up the sides of grooves, rebates and housings
Side rebate plane	Fine cleaning up and finishing both with grain and for end grain.
Block plane	Used to clean out the waste wood from housings and grooves
Bull nose plane	Has cutting iron mounted towards the front for working stopped rebates and chamfers
Shoulder plane	Used for the final working and cleaning up of curved edges

  ACTIVITY

1) State where a bench plane with a corrugated sole would be better to use than a standard smooth-sole one.

2) List the procedure to be followed when preparing sawn timber to size by hand.

Your questions answered...

My bullnose plane has a removable front end. What's the purpose of this and has it a special name?

The front end is removed to enable closer working into stopped rebates and stopped chamfers. Bullnose and shoulder planes with a removable front end are often termed as 'chisel planes'.

Use carpentry and joinery hand-held drills

Hand-held drills

Woodworkers traditionally used bradawls, hand-drills and ratchet braces to bore holes in timber for a variety of purposes. However, with the increasing availability of power tools and especially battery-powered drills, their use is far less commonplace today.

ACTIVITY

The most difficult part of drilling or boring is to start and keep the drill or bit going in the right direction.

1) Describe TWO practical measures that can be taken to ensure that a drill or bit is not leaning out of line when boring.

2) Explain how breaking out can be avoided when boring holes through a timber section with a brace and bit.

3) Name a type of twist drill that has a feature to prevent it slipping off the mark when starting to drill in timber.

Drill bits

Drill bits for use in a ratchet brace are often termed auger bits. They have a square tang on the end to hold them secure in the chuck.

ACTIVITY

Look at the illustrations, which show a range of bits that are used in a ratchet brace. Name and state a typical use for each.

A)

B)

C)

D)

E)

F)

G)

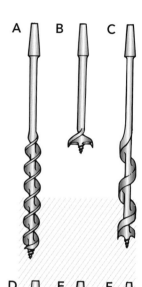

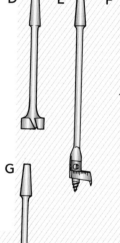

Hand-held drills

ACTIVITY

1) State the alternative name for a hand drill that is used with twist drill bits.

2) HSS twist drills are commonly used to form holes in wood, metal and plastic. What does HSS stand for?

3) Describe a situation where the ratchet on a carpenter's brace would be put into action.

Woodworking chisels

There are many types of chisels, each having a specific purpose: cutting, chopping and cleaning up joints; forming housings and recesses to accommodate ironmongery: paring and shaping timber rather than using a saw or plane.

A

B

Thicker rectangular blade

C

D

Longer lighter blade

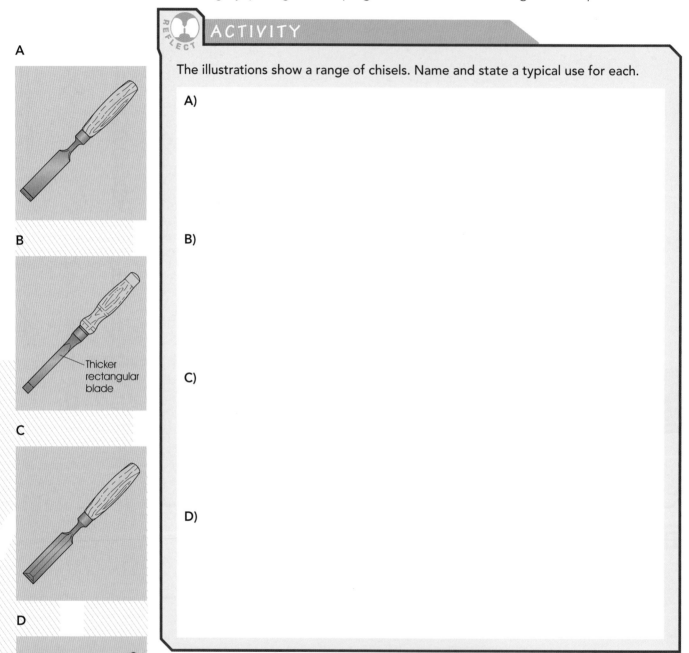

ACTIVITY

The illustrations show a range of chisels. Name and state a typical use for each.

A)

B)

C)

D)

Your questions answered...

What is 'paring'?

Paring refers to the use of a chisel to cut or remove thin slices of wood at right angles to the grain or across the end grain, normally using hand pressure only.

Chisel sizes and blade sections

Chisel sizes are determined by the width of the blade, which range from 3 mm up to 50 mm. Most woodworkers will start with a basic set of bevelled-edge chisels consisting of 6 mm, 9 mm, 12mm, 18 mm and 25 mm widths, adding other widths and sections later to suit the tasks regularly undertaken.

ACTIVITY

Produce sketches to show the cross-section of the following chisel blades:

Firmer

Mortise

Bevelled

Gouge

Woodworking chisels

1) Explain the difference between a paring chisel and a bevelled-edge chisel.

2) Describe the procedure used to chop a mortise.

3) State how a chisel should be safely passed from one person to another.

4) State why chisel handles may be fitted with steel ferrules or caps.

5) Explain the difference between firmer and scribing gouges.

Risks of leaving tools unattended

In a small group or on your own, consider the following real-life problem:

Jude is a recently qualified carpenter who works for a local jobbing builder that specialises in the maintenance of public and commercial buildings.

Her latest job was to adjust and replace the ironmongery on a pair of doors for the side entrance of a junior school.

Before going to lunch, Jude collected her tools and equipment together and placed them under a dust sheet just outside the doors.

On returning to work after lunch, it was apparent there had been a short rain shower, leaving the dust sheet wet and the tools underneath damp.

1) What should Jude do to prevent the tools from rusting?

2) Were there any other risks involved in this situation?

3) What actions do you think Jude should take in the future?

Carpentry and joinery hand tools

ACTIVITY

Solve the following clues and find these words hidden in the word square. You may find the words written forwards, backwards, up, down or diagonally.

1) Legislation that covers the supply and use of grinding wheels (abbreviation). (5)
2) A grinding wheel that is lubricated with water. (9)
3) A type of drill bit used to bore flat-bottomed holes. (8)
4) A chisel having a curved section blade. (5)
5) The longest type of bench plane. (3)
6) A type of firmer chisel which has a steel band at the end of its handle. (10)
7) The process of bending over the teeth of a saw blade. (7)
8) The width of a saw cut. (4)
9) The general name for tenon, dovetail and gents' saws. (8)
10) A saw used to cut metal and plastics. (7)
11) A type of plane used to form grooves. (6)
12) The bevel applied to plane and chisel blades after grinding. (6)
13) A teardrop section oil or waterstone used when sharpening a gouge. (9)
14) A small plane used for the fine trimming of end grain. (5)
15) A plane used to finish stopped rebates and chamfers. (8)
16) The word represented by the middle letter in the abbreviation HSS. (5)
17) The shortest type of bench plane. (9)
18) The base of a plane. (4)
19) To saw timber to width. (3)
20) A narrow blade saw used to cut keyholes. (6)

F	J	R	J	A	X	H	E	D	C	E	G	U	O	G	M	C	W
E	E	C	N	R	N	K	A	H	Q	H	Q	B	N	X	D	T	F
D	Y	S	U	I	E	O	O	I	T	F	O	I	S	S	C	C	A
W	W	W	Y	R	T	R	F	O	R	S	T	N	E	R	T	L	E
R	H	A	F	E	T	I	D	P	M	T	B	V	I	O	W	H	R
A	S	S	R	G	U	T	U	L	E	U	B	S	G	N	O	C	G
R	M	K	O	I	N	I	N	S	L	M	S	G	D	O	G	B	N
H	A	C	K	S	A	W	E	L	A	N	S	Q	N	O	T	R	U
W	M	A	P	T	I	T	N	I	O	N	S	C	F	E	P	U	M
C	C	B	P	E	M	O	A	P	N	S	D	N	E	V	U	V	F
N	Y	F	O	R	S	S	A	S	E	K	V	S	A	X	W	C	I
T	Q	M	D	E	E	S	Y	T	C	A	R	R	T	Q	E	E	M
U	Q	A	E	D	W	O	P	O	D	H	H	N	V	O	R	I	P
M	E	C	T	S	A	B	L	N	Y	P	A	L	I	Z	N	D	A
D	V	K	V	O	D	B	O	E	R	R	Z	G	C	X	E	E	D
W	B	N	S	R	G	F	U	U	A	O	Q	W	Y	E	L	O	S
N	B	E	J	S	C	S	G	U	G	P	V	Q	P	B	B	W	A
N	B	Q	Y	G	N	I	H	T	O	O	M	S	V	B	D	G	W

QUICK QUIZ

1. Which of the following planes is most suitable to use when forming grooves for panels?
   a. plough
   b. rebate
   c. router
   d. shoulder

2. The main purpose of a back iron, which is used in bench planes, is to:
   a. stiffen the cutting iron.
   b. secure the cutting iron.
   c. break the shavings.
   d. protect the cutting edge.

3. The most suitable plane to use when preparing a long straight edge is a:
   a. block plane.
   b. try plane.
   c. smoothing plane.
   d. jack plane.

4. The correct grinding and honing angles to use when sharpening plane and chisel blades are:
   a. grinding 20° honing 25°.
   b. grinding 25° honing 30°.
   c. grinding 25° honing 20°.
   d. grinding 30° honing 25°.

5. In the absence of a mortise chisel, which of the following would be the most suitable for chopping mortise joints?
   a. paring
   b. bevelled edge
   c. firmer
   d. gouge

6. The tool most suited to easing the sides of a panel groove is a:
   a. paring chisel.
   b. rebate plane.
   c. plough plane.
   d. side rebate plane.

7. The process of filing the cutting edge of a saw to bring all of the teeth in line is:
   a. topping.
   b. dressing.
   c. shaping.
   d. setting.

8. A tenon saw is commonly used in the workshop to cut:
   a. tenons and other joints along the grain.
   b. fine mouldings and sheet material.
   c. joint shoulders and general crossing cutting on the bench.
   d. dovetails, small mouldings and other delicate work.

9. A common piece of equipment used when cross-cutting on a bench is a:
   a. shooting board.
   b. bench hook.
   c. box square.
   d. saw horse.

10. Which one of the following is not normally used with a ratchet brace?
    a. twist drill bit
    b. centre bit
    c. auger bit
    d. expanding bit

UNIT 1006

Prepare and use carpentry and joinery portable power tools

Woodworkers have at their disposal a wide range of powered hand tools, enabling many operations to be carried out with increased speed, efficiency and accuracy. However, they must undertake training to be competent to use them and be aware of all the safety aspects associated with their use.

Key knowledge:
- ➤ maintenance and storage of portable power tools.
- ➤ using portable power tools to cut, shape and finish.
- ➤ using portable power tools to drill and insert fastenings.

Maintain and store portable power tools

Symbols and labels

ACTIVITY

Use the internet to look up the following symbols and labels, which are associated with portable power tools. Explain what each means and name a type of portable power tool they would be displayed on.

1) CE mark.

2) BS kitemark.

3) Double insulation mark.

4) PAT pass certificate .

Legislation associated with portable power tools

ACTIVITY

The table details the main requirements of different acts and regulations that are associated with portable power tools. Complete the table to show (1) the name and (2) common abbreviation of each regulation or act described by the requirements.

Requirement of legislation	Name and abbreviation
Requires employers to provide employees with any necessary personal protective equipment that is required in order to carry out work safely.	1) 2)
Requires employers to control exposure to hazardous substances in the workplace to prevent ill health and protect both employees and others who may be exposed.	1) 2)
The main statutory legislation that covers the health and safety of all persons at their place of work, and other people from the risks occurring through work activities.	1) 2)
Requires the prevention or control of risks to people's health and safety from equipment they use at work.	1) 2)

What safety equipment is required?

ACTIVITY

The tasks shown in the table are carried out, on-site, by various workers using portable power tools.

Either on your own or in a small group, consider the tasks listed, and complete the table to show what items of safety equipment should be used for each.

Task	Safety equipment required
James is going to use a 110 V hammer-action drill from a 230 V socket to drill holes in brickwork for fixing bolts.	
Jade is using a power router to mould the edge of some MDF strips.	
Rashid is going to fix timber battens to clad concrete columns and has been told to use a cartridge-operated fixing tool.	
David and Jude are using gas-operated nail guns to fix fascia boards to the eaves of a roof.	

Maintenance and storage of portable power tools

ACTIVITY

1) Which power source is considered to be the safest for powered hand tools, especially in damp conditions?

2) What action would you take if you suspected a portable power tool was not working correctly or was unsafe?

3) Where can information concerning the use and maintenance of a particular power tool be found?

4) Why is the use of 110 V power tools considered to be safer than use of 230 V power tools?

5) A 110 V step-down transformer is to be used from a 230 V mains supply socket. Where, if required, should an extension lead be positioned?

Use portable power tools to cut, shape and finish

Identification of power tools used for cutting, shaping and finishing

ACTIVITY

The illustrations show a range of portable power tools.

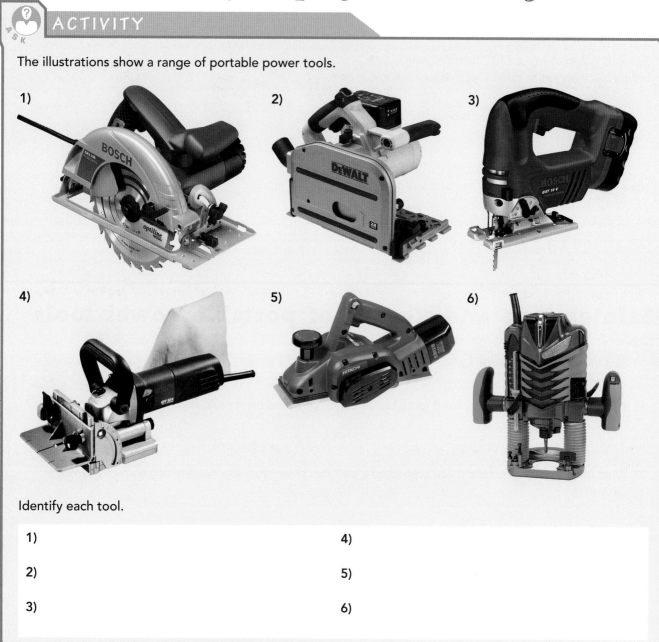

1)

2)

3)

4)

5)

6)

Identify each tool.

1)

2)

3)

4)

5)

6)

Your questions answered...

Why have I been instructed to fully unwind an extension lead from its drum before use?

If an extension lead is left partly coiled in use, it will start to heat up when supplying power, creating a fire risk. You must also make sure the uncoiled lead is carefully routed so as not to create a tripping hazard.

Router cutters

ACTIVITY

The images show router cutters.

1) Produce sketches under each of the cutters to show the shape of mould they would produce. Name each shape.

A) 　　B) 　　C) 　　D) 　　E)

ovolo

eg.

2) Explain what HSS and TCT mean in relation to router cutters. State the advantage of using one over the other.

HSS:
TCT:
Advantage:

3) Produce a sketch in the space below to show a typical router cutter that could be used to trim the edges of laminate or timber lippings.

Using portable power tools to cut, shape and finish

ACTIVITY

1) Describe how a portable power planer may be used to form a small chamfer on the edge of a timber section.

2) State how the area surrounding the use of portable powered planers, sanders and routers can be protected from any fine dust or shavings that may be produced.

3) Name TWO portable power tools that may be used to form rebates on a timber section.

   A)

   B)

4) Abrasive papers are graded according to the size of grit. Which grit size will give the finest finish: 80grit; 120grit; or 180grit?

Using portable power tools to drill and insert fastenings

Identifying power tools used for drilling and inserting fastenings

ACTIVITY

Name each tool and state its typical use.

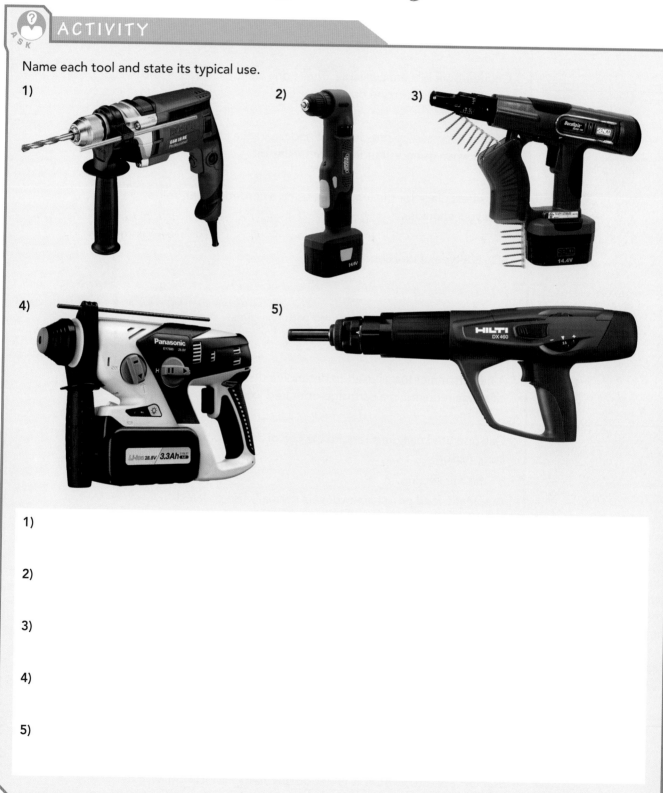

1)

2)

3)

4)

5)

1)

2)

3)

4)

5)

Drills, screwdrivers and nail guns

THINK ACTIVITY

Tick whether the following statements are true or false.

Statement	True	False
A fast speed should be selected when inserting and removing screws.		
Counter-bore bits are available. These drill a pilot, clearance and counter-bore hole in one operation.		
It is not necessary to pre-drill pilot and clearance holes for screws when using a standard countersink bit.		
The safest place for the lead when using a corded power drill is over your shoulder.		
SDS is a type of drill chuck.		
BPM is an abbreviation associated with nail guns.		
A hand hammer test is used to determine the strength of cartridge required in a cartridge-operated fixing tool.		
Gas-operated nail guns require the use of an independent air compressor.		
Automatic-feed power screwdrivers normally use collated screws.		
Drill/drivers with an adjustable torque control should be set to the maximum when drilling.		

Your questions answered...

Do battery-powered hand tools require a PAT test?

No, the tool does not require a PAT test, but the battery charger will. However, the battery-powered tool itself still requires regular inspection and maintenance to ensure it is in good working condition.

Use portable power tools to drill and insert fastenings

ACTIVITY

List FOUR safety precautions to be followed when using portable power tools.

1)

2)

3)

4)

ACTIVITY

1) State the purpose of a 'torque' control on a battery-operated drill/driver.

2) The strength of cartridge to be used in a cartridge-operated fixing tool is indicated by a colour code. Circle the code which represent the strongest cartridge.

Red   Yellow   **Brown**

3) What might the result be if contact is not maintained between a gas-operated nail gun and the fixing surface when the trigger is operated?

4) Explain what can be done to minimise the risk of coming into contact with hidden service runs when drilling into walls.

Preparing and using portable power tools

 ACTIVITY

Find the following 20 words, all associated with portable power tools, in the word square. You may find the words written forwards, backwards, up, down or diagonally. You will need to become familiar with all of these terms.

Orbital	Recessing	Counterbore
Compressed	Mitre saw	Pneumatic
Plunge	Chuck	Surfacing
Specification	Drill	Template
Jigaw	Router	Collated
Twistdrill	Electric	Torque
Screwdriver	Extraction	

A	C	T	D	M	I	N	E	L	E	C	T	N	M	K	N	O	P
B	T	W	E	U	O	A	O	E	M	I	O	U	C	T	J	L	H
R	E	I	S	R	P	R	Y	I	I	I	S	U	A	I	L	A	E
E	N	S	S	E	E	T	H	I	T	C	H	N	G	M	A	T	M
V	S	T	E	L	L	C	W	A	R	C	X	S	Y	N	E	I	A
I	I	D	R	I	L	L	C	E	E	H	A	G	N	M	R	B	T
R	O	R	P	R	E	I	W	E	S	W	I	R	P	P	O	R	I
D	N	I	M	D	F	E	J	K	A	L	M	L	T	A	B	O	C
W	N	L	O	I	U	G	F	D	W	E	A	T	L	X	R	O	B
E	O	L	C	K	A	N	B	I	R	T	G	R	T	I	E	T	E
R	B	E	H	T	A	U	E	S	E	U	A	C	E	B	T	Y	L
C	P	A	V	O	E	L	A	B	C	I	G	T	H	T	N	R	E
S	K	F	F	R	I	P	T	S	E	G	N	I	B	B	U	U	C
N	O	B	P	Q	R	A	T	A	S	R	E	F	I	N	O	O	T
O	I	F	Y	U	O	C	A	N	S	E	E	T	H	I	C	S	R
G	T	P	N	E	U	M	A	T	I	C	O	N	I	S	I	T	I
I	O	C	K	O	N	T	K	Y	N	D	E	T	A	L	L	O	C
N	S	U	R	F	A	C	I	N	G	A	E	O	S	T	A	N	P

1. All portable electrical appliances have to be tested by a competent person at regular intervals to ensure they are safe to use. This is known as a:
   a. PUWUR test.
   b. PAT test.
   c. HSS test.
   d. COSHH test.

2. The main hazard to health when drilling walls using a hammer-action drill is:
   a. heat exhaustion.
   b. electrocution.
   c. dust.
   d. vibration.

3. The best portable power tool to use when cutting curves in timber-based manufactured boards is a:
   a. circular saw.
   b. chop saw.
   c. jig saw.
   d. biscuit jointer.

4. The chuck on a handheld portable power drill is best described as:
   a. the part where drill and screwdriver bits are inserted and held securely.
   b. the part adjacent to the trigger, which is pushed to allow continuous drilling.
   c. the part near the motor that adjusts the torque to allow screws to be driven flush with the surface.
   d. the part where screws and nails are inserted and held securely, ready for driving in.

5. What is the main reason 110V power tools must be used on building sites rather than 230V power tools?
   a. 110V power supply is cheaper.
   b. 110V power supply is safer.
   c. 110V power tools are quicker to use.
   d. 10V power tools are more readily available.

6. What is the best way to deal with the dust and chips created when using portable, powered routers?
   a. Wear a dust mask and face shield.
   b. Stop work and sweep up periodically.
   c. Connect the router to a dust extractor.
   d. Sweep up after the work is finished.

7. Which one of the follow sanders is best for fine-finishing installed window boards?
   a. belt sander
   b. delta sander
   c. palm sander
   d. orbital sander

8. When using an extension lead with a portable power tool, it should be:
   a. fully unwound from the drum.
   b. coiled on the drum.
   c. coiled on the floor.
   d. kept as short as possible.

9. The recommended items of personal protective equipment to be worn when using a gas-operated nail gun are:
   a. safety helmet, dust mask and ear protectors.
   b. safety helmet, eye protection and ear protectors.
   c. dust mask, eye protection and ear protectors.
   d. safety helmet, eye protection and dust mask.

10. The portable power tool that can be used with a jig to cut dovetail joints is a:
    a. biscuit jointer.
    b. router.
    c. plunge saw.
    d. jig saw.

Glossary

Absolute: means that the requirement must be met regardless of cost or other implications.

Approved codes of practice (ACoP): documents published from time to time, which contain guidance, examples of good practice and explanations of the law.

Area: the extent or measurement of a surface.

Building: the act or process of creating and maintaining the built environment.

Building information: the key contact documents produced by the clients, such as the working drawings, bill of quantities, specification, schedules, specification and conditions of contract. In addition it also refers to the work programmes produced by a contractor and any manufacturers' information.

Buildings: structures that enclose space and in doing so create an internal environment. The actual structure of a building is termed the external envelope, which protects the internal environment from the outside elements, known as the external environment.

Communicating: the exchange of thoughts, messages and information, by speech, body language, written or printed information and telecommunications.

Construction: the process of building or assembling the built environment.

Duty: something that a person or organisation is expected or required to do.

Elements: the constructional parts of a building's substructure and superstructure. They are classified into three main groups: primary elements, secondary elements and finishing elements, depending on their function.

Floors: The horizontal ground and upper levels in a building that provide a walking surface.

Formulae: mathematical relationships or rules expressed using symbols.

Foundations: the part of a structure that transfers the loads of the superstructure safely onto the supporting ground.

Handrail: a rail that is designed to provide a person with stability and support when using stairs and associated landings.

Harm: can vary in severity, some hazards can cause death, other illness or disability or maybe only cuts or bruises.

Hazard: something with the potential to cause harm.

Joist: one of a series of parallel beams that span the gap between walls in suspended floors and roofs, to support floor, ceiling and flat roof surfaces.

Legislation: a law or set of laws written and passed by Parliament, which is often referred to as an Act of Parliament.

Maintenance: tool maintenance is the regular upkeep and sharpening of tools to ensure they remain in good condition.

Perimeter: the extent of the boundaries of a shape.

Reasonably practicable: means that you are required to consider the risks involved in undertaking a particular work activity. However, if the risks are minimal and the cost or technical difficulties of taking certain actions to eliminate the risks are high, it might not be reasonably practicable to implement those actions.

Regulations: orders issued by a government department or agency that has the force of law.

Risk: concerned with the severity of harm and the likelihood of it happening.

Roof: the uppermost part of a building, that spans the walls and protects the building and its contents from the effect of weather.

Substructure: the structure from below the ground up to and including the ground floor. Its purpose is to receive the loads from the main building superstructure and its contents, and transfer them safely down to a suitable load-bearing layer of ground.

Superstructure: all of the structure above the substructure, both internally and externally. Its purpose is to enclose and divide space, and transfer loads safely to the substructure.

Walls: the vertical enclosing and dividing elements of a building.